D1453580

PERCHLORATE
Environmental Problems and Solutions

PERCHLORATE

Environmental Problems and Solutions

Kathleen Sellers
William Alsop
Stephen Clough
Marilyn Hoyt
Barbara Pugh
Joseph Robb
Katherine Weeks

Taylor & Francis
Taylor & Francis Group
Boca Raton   London   New York

CRC is an imprint of the Taylor & Francis Group,
an informa business

CRC Press
Taylor & Francis Group
6000 Broken Sound Parkway NW, Suite 300
Boca Raton, FL 33487-2742

© 2007 by Taylor & Francis Group, LLC
CRC Press is an imprint of Taylor & Francis Group, an Informa business

No claim to original U.S. Government works
Printed in the United States of America on acid-free paper
10 9 8 7 6 5 4 3 2 1

International Standard Book Number-10: 0-8493-8081-2 (Hardcover)
International Standard Book Number-13: 978-0-8493-8081-5 (Hardcover)

Library of Congress Cataloging-in-Publication Data

Perchlorate : environmental problems and solutions / by Kathleen Sellers ... [et al.].
    p. cm.
Includes bibliographical references and index.
ISBN 0-8493-8081-2 (alk. paper)
    1. Perchlorates--Environmental aspects. 2. Water--Purification--Perchlorate removal. 3. Soil pollution. 4. Soil remediation. I. Sellers, Kathleen. II. Title.

TD427.P33.P4285 2006 2007
628.5'5--dc22
                                                  2006047552

Visit the Taylor & Francis Web site at
http://www.taylorandfrancis.com

and the CRC Press Web site at
http://www.crcpress.com

The Authors

The AMEC team in Westford, Massachusetts has worked together to solve perchlorate contamination problems by integrating their knowledge of contaminant fate and transport, measurement, risk assessment, and remediation.

Kathleen Sellers, PE

Ms. Sellers has focused on developing and designing solutions to hazardous waste and wastewater treatment problems. She has worked on a variety of sites including perchlorate releases, former chemical manufacturing facilities, manufactured gas plants, petroleum releases, and landfills. Her work, under both state and federal regulatory programs, has included development of site investigation plans, management of focused field investigations, feasibility studies, remedial design and installation, negotiations with regulatory agencies, and public involvement programs. Ms. Sellers published an engineering textbook entitled *Fundamentals of Hazardous Waste Site Remediation* (CRC Press/Lewis Publishers, 1999).

Katherine R. Weeks, PE

Ms. Weeks has specialized in evaluating and designing treatment technologies for perchlorate and explosives residuals in groundwater. She recently managed the design and construction of mobile *ex situ* treatment systems for the removal of perchlorate and explosives from groundwater that can utilize either granular activated carbon or ion exchange resins as the treatment medium. She has also incorporated a biological fluidized bed reactor (designed by others) into a design for an *ex situ* treatment system for the removal of perchlorate and groundwater. Ms. Weeks has managed oversight of fourteen bench- and pilot-scale studies for remediation of perchlorate and explosives in soil and groundwater using innovative technologies. After 7 years with AMEC, Ms. Weeks has recently accepted a position as Senior Engineer with the Town of Framingham in Massachusetts.

William R. Alsop

Mr. Alsop has more than twenty years' experience in human health and ecological risk assessment projects. He has evaluated the ecological effects of direct and food chain exposures to perchlorates, explosives residuals, dioxins and furans, chlorinated benzenes, metals, pesticides, polychlorinated biphenyls (PCBs), and polynuclear aromatic hydrocarbons (PAHs). Mr. Alsop co-authored a chapter entitled *Ecological Risk Assessment Applied to Energy Development* in the *Encyclopedia of Energy* (C.J. Cleveland, ed., Elsevier, 2004).

Stephen R. Clough, Ph.D., DABT

Dr. Clough has over twenty years of experience in environmental consulting. He has diverse expertise in the areas of ecological and human health risk assessment, aquatic toxicology, water quality and environmental modeling. He has participated in and managed numerous ecological risk assessment projects for several private and public organizations, including work on perchlorate-contaminated sites. As a board-certified toxicologist, Dr. Clough is often called to help support litigation efforts and to appear as an expert witness.

Marilyn Hoyt

Ms. Hoyt has over twenty years of experience in environmental consulting with a particular focus on programs requiring expertise in chemistry and quality assurance. She has managed studies requiring original research with integration of literature reports and supported health risk studies requiring data collection and interpretation. Ms. Hoyt has directed a full-service environmental laboratory and is familiar with the current and developing analytical techniques for analyses of perchlorate. She has presented information on perchlorate and other environmental methods and data interpretation to public agencies and citizen groups. Her work has included measurement program design and implementation for hazardous waste site investigations and remediation projects, preparation of Quality Assurance Plans and negotiation with EPA and state agency staff on final requirements. Ms. Hoyt has served as an expert witness and provided technical preparation for parties involved in litigation.

Barbara Pugh

Ms. Pugh has over ten years of experience in environmental consulting. She has managed and conducted human health risk assessments at both state and federal regulated CERCLA and RCRA hazardous waste sites, including military facilities, petroleum refineries, industrial/manufacturing facilities, former manufactured gas plants, and landfills. Her experience includes evaluating human health risks associated with exposures to explosive compounds and related compounds (including perchlorate), petroleum and polyaromatic hydrocarbons, chlorinated solvents, polychlorinated biphenyls (PCBs), pesticides, herbicides, dioxins, and metals. In addition, Ms. Pugh is experienced in providing support negotiating regulatory, guidance and policy interpretations in matters related to risk assessment, risk management, and clean-up levels.

Joseph Robb, P.G.

Mr. Robb has over nine years of experience in hazardous waste site characterization and remediation and has published technical articles on multiple occasions. His work has focused on evaluating the nature and extent of environmental contamination, hydrogeologic characterization, natural attenuation assessments, and subsurface remedial system design, operation and maintenance. His experience includes sites contaminated with chlorinated hydrocarbons, petroleum hydrocarbons, heavy metals, coal tar, explosives, propellants and perchlorate. Mr. Robb uses an understanding of the physical, chemical and biological processes that control the fate and transport of contaminants in the subsurface to characterize and remediate hazardous waste sites. Mr. Robb has considerable experience with evaluating military ranges and training areas for the presence and potential for migration of perchlorate. Mr. Robb recently used fate and transport modeling to develop soil clean-up goals for perchlorate at a military installation.

Acknowledgments

The authors gratefully acknowledge the scientists and engineers who reviewed drafts of this manuscript and provided valuable perspective:

Jacimaria R. Batista, Ph.D. - Department of Civil and
    Environmental Engineering, UNLV
Jay L. Clausen, CPG - U.S. Army Corps of Engineers
Bill Gallagher - Army National Guard
Marc Grant, PE - AMEC Earth and Environmental
Bruce Hope, Ph.D. - Oregon Department of Environmental Quality
Ian Osgerby, Ph.D., PE - U.S. Army Corps of Engineers
Kevin H. Reinert, Ph.D. - AMEC Earth and Environmental
Scott Veenstra - AMEC Earth and Environmental
Randall Wilkinson - Army National Guard

Robin Overbaugh, supported by Ronald W. Bowman, Kate Dickson, and Alisa Planson of AMEC, provided graphic design services and assisted with preparation of the manuscript.

Dedication

This book is dedicated to Bradley W. Schwab (1950-2004) Ph.D., D.A.B.T.: mentor, friend, and inspiration.

Table of Contents

CHAPTER 1

Introduction

Prized for its ability to power rocket engines and used in many industrial processes, perchlorate now garners headlines for concerns over environmental pollution. The U.S. Environmental Protection Agency (U.S. EPA) estimates that perchlorate affects the drinking water of over 11 million people. When chemists developed analytical methods which identified widespread perchlorate contamination in the late 1990s, research into the environmental problems and potential solutions exploded. The purpose of this book is to summarize the state of the science and developments in engineering relative to perchlorate in the environment. This book describes

- Common sources of perchlorate,
- Its behavior in the environment,
- Methods for analyzing perchlorate in environmental samples,
- Potential risks to human health and the environment,
- Regulatory standards and criteria, and
- Techniques for remediating environmental contamination.

It illustrates these points with case studies of sites where perchlorate contaminates soil, groundwater, and surface water. These case studies provide perspective on issues commonly faced by scientists, engineers, and managers of perchlorate-impacted sites.

The remainder of this preface provides an overview of critical information. Subsequent chapters provide more detail and the literature citations for the information presented in this preface.

1.1 What Is Perchlorate?

Perchlorate is an ion containing chlorine and oxygen, with a chemical structure abbreviated ClO_4^-. In general, the perchlorate ion is highly soluble, nonvolatile, and stable under most environmental conditions. As discussed further in subsequent chapters, the physical and chemical properties of perchlorate compounds determine their behavior in the environment and their toxicity. These properties also form the basis for analytical methods and techniques for remediating environmental contamination.

Perchlorate can originate from natural sources, which include some soils in arid climates derived from ancient marine seabeds. Some of those soils have been incorporated into fertilizers used in the United States on crops such as tobacco, cotton, and some fruits. Perchlorate may be created in very low concentrations during lightning storms and subsequently deposited onto surface soils.

Manufacturers produce six perchlorate compounds in large amounts: magnesium perchlorate, potassium perchlorate, ammonium perchlorate, sodium perchlorate, lithium perchlorate, and perchloric acid. Defense activities and the aerospace industry use approximately 90% of perchlorate products, primarily ammonium perchlorate. While perchlorate's major uses are in the production of munitions, explosives and fireworks, manufacturers also use perchlorate compounds in small amounts in some high-volume consumer products. Since 1976, over 14,000 patents have been issued for the use of various perchlorate-containing materials.

Many of the sites contaminated with perchlorate were once associated with the manufacture of perchlorate or its use in defense-related operations such as rocket manufacture or munitions use or demolition. Other environmental contamination has resulted from the use of perchlorate in applications such as fireworks, flares, or blasting agents.

In addition to the natural and anthropogenic perchlorate sources discussed above, perchlorate can be found as a breakdown product of sodium hypochlorite (i.e., bleach) and can be incidentally formed in corrosion control applications.

1.2 How Is It Measured?

Since perchlorate was not recognized as a contaminant of general concern in the environment until relatively recently, U.S. EPA test methods did not address perchlorate until the late 1990s. The scientific literature reported a variety of analysis methods for perchlorate as far back as the early 1900s. While these methods were relatively straightforward, most were designed more for bulk analyses of materials or analyses of relatively high concentration solutions, and their limited sensitivity and selectivity do not suffice for trace level analyses of perchlorate in environmental samples.

The period between 1999 and 2005 saw a high level of analytical perchlorate research, with over 100 presentations and publications on method development. Investigators can now use a variety of both field and laboratory analyses to detect perchlorate in environmental samples. This rapid development and regulatory approval of analytical methods places those needing perchlorate analyses in the position of having a complicated menu of choices, each with advantages and limitations or disadvantages. Understanding the methods may be critical to obtaining data that will support subsequent uses for regulatory submittals, site characterizations, source attribution, remedial feasibility studies, risk assessment and ongoing monitoring. Each of these data usages has specific requirements, and a method appropriate for one use may not meet the needs of other requirements. A less sensitive method may not provide data required to demonstrate the absence of risk, but a method that has very low detection limits may be less accurate and more costly for samples with high concentrations.

Laboratory methods generally rely on ion chromatography or liquid chromatography to separate perchlorate ions from solution, then measure perchlorate concentrations using conductivity detection or mass spectrometry. Because perchlorate is an ion, chemists can measure the amount of perchlorate in solution using conductivity detection. However, that measurement may also reflect other ions in solution and is therefore considered a "presumptive analysis" for perchlorate. Mass spectrometry can provide definitive identification of perchlorate, however interferences remain a concern and may result in false negatives or elevated reporting limits.

Chemists have developed forensic analyses to distinguish whether perchlorate in an environmental sample originated from natural material, such as Chilean nitrate fertilizer or a man-made material. During isotopic analysis of the chloride released from natural and manufactured sources, researchers noted consistent differences in the $^{37}Cl/^{35}Cl$ and $^{18}O/^{17}O/^{16}O$ isotope relationships that result from the formation mechanisms. This finding has since formed the basis for source attribution of perchlorate in several instances.

Forensic analyses can also indicate the progress of perchlorate biodegradation. Laboratory studies have found that chlorine isotope analysis can document perchlorate biodegradation and potentially can distinguish this process from other non-biological mechanisms that reduce perchlorate levels during *in situ* remediation efforts. The perchlorate-reducing bacterium, *Azospira*, preferentially reduces perchlorate anions with the ^{35}Cl isotope. Thus the relative abundance of the $^{35}Cl:^{37}Cl$ ratio in water can indicate microbial degradation as a removal pathway.

Researchers are continuing to develop analytical methods for use in both the laboratory and the field.

1.3 How Does Perchlorate Move through the Environment?

Perchlorate is highly soluble and stable under typical environmental conditions. As a result, it can move readily through the environment when dissolved in migrating soil pore water, groundwater, and surface water. Perchlorate salts do not migrate through the vapor phase due to their low vapor pressure. Once released to the environment, the physical, chemical, and biological processes that affect the fate and transport of perchlorate include dissolution of source material (in the case of solids), advection, dispersion and diffusion, sorption, and biological degradation under anaerobic conditions.

1.4 What Are the Implications of Human and Ecological Exposures to Perchlorate?

Scientists assessing the possible risks to humans from chemical exposures consider different types of information about the chemical, including possible sources in environmental media, the toxicity of the chemical and whether the chemical is considered a human carcinogen, and the conditions of possible exposure. Much of the concern regarding the toxicity of perchlorate has stemmed from its widespread occurrence in drinking water. Current estimates suggest that public water supplies for over 11 million people in the United States contain detectable perchlorate. Most of these detections, however, are less than 12 micrograms per liter (μg/L). Exposure to low levels of perchlorate can competitively inhibit the uptake of iodide by the thyroid gland. This effect is reversible, and one would have to reduce iodide uptake by at least 75% for several months or longer before experiencing a decline in thyroid hormone production that would have adverse health effects. No evidence suggests that perchlorate causes thyroid disorders, thyroid nodules or cancer in the thyroid gland or any other organ. It is also not genotoxic or mutagenic.

Ecological risk assessment differs in a fundamental respect from human health risk assessment. Rather than considering effects on a population of a single species - that is, humans - ecological risk assessors must evaluate the risks to ecosystems containing perhaps thousands of interdependent species. The process parallels human health risk assessment in its consideration of exposures and potential health effects. Studies of aquatic and terrestrial organisms show that, due to its solubility, perchlorate generally passes readily through the organism and does not bioaccumulate. Exposure to perchlorate can affect iodide uptake in animals that have thyroid glands. At high concentrations, perchlorate can affect the endocrine system, disrupting seed germination in plants and affecting development in amphibians during sensitive developmental life stages. These effects are reversible, however, and their significance depends in part on whether they affect the viability of the ecological community.

1.5 What Are the Regulatory Limits on Perchlorate?

Decision-makers use the information from a risk assessment to develop strategies to minimize or eliminate risks due to chemical exposures. Such risk management decisions include the determination of regulatory limits. Two types of limits are particularly important for perchlorate: Maximum Contaminant Limits on perchlorate in drinking water, determined under the Safe Drinking Water Act, and Ambient Water Quality Criteria developed under the Clean Water Act.

At the federal level, the Safe Drinking Water Act authorizes the U.S. EPA to set enforceable drinking water standards, also referred to as Maximum Contaminant Limits (MCLs). Individual state environmental agencies may choose to develop their own drinking water standard or adopt the federal MCL. In general, the U.S. EPA derives drinking water standards for contaminants found in public water supplies at concentrations that pose a likely public health concern. For perchlorate, several issues have had varying impacts on the development of a MCL, including the availability of analytical methods to truly determine if perchlorate is present, the determination of a significant risk, and suitable treatment options.

In 1998, the U.S. EPA added perchlorate to the Drinking Water Contaminant Candidate List (CCL). On February 24, 2005, the EPA included perchlorate on the second CCL. This designation provides for comprehensive studies regarding analytical methods for detecting the contaminant, the prevalence of occurrence in drinking water, potential health effects, and the efficacy of treatment technologies to remove the contaminant from drinking water. Based on the information from these studies, the U.S. EPA will formally determine by August 2006 whether it should issue a national primary drinking water regulation for perchlorate.

The process of setting a MCL begins with determining a dose-response factor known as a reference dose. Based on the potential health effects and allowing a margin of safety, the U.S. EPA defined a reference dose of 0.0007 milligrams per kilogram per day (mg/kg•day). This reference dose corresponds to a Drinking Water Equivalent Level, which is a guideline for the maximum allowable concentration of perchlorate in drinking water assuming no other consumption of perchlorate, of 24.5 micrograms per liter (μg/L) for adults.

When no drinking water standard has been promulgated, a health advisory (HA) represents a non-enforceable guideline set by federal or state regulators to evaluate the health significance of a contaminant in drinking water. Nine states have issued health advisories for perchlorate as of December 2005, ranging from 1 to 18 μg/L.

Acute and chronic Ambient Water Quality Criteria (AWQC), developed under the Clean Water Act, are the U.S. EPA's benchmarks for evaluating aquatic toxicity. While regulators use AWQC to develop wastewater discharge permit limits, for example, they are not promulgated regulatory limits in and of themselves. The U.S. EPA develops AWQC values through a formal protocol which includes data from a series of toxicity tests.

The U.S. EPA has used bioassay data to calculate acute and chronic AWQC for perchlorate of 22.3 milligrams per liter (mg/L) and 10.3 mg/L, respectively, but cautioned that because these benchmarks "have not been promulgated by the Office of Water nor have undergone full peer review, they cannot be considered national ambient water quality criteria at this time." Due to concerns that the chronic AWQC generated using the standard protocol might not be sufficiently protective of sensitive life stages of amphibians, U.S. EPA used data on the effect of perchlorate on the development and metamorphosis of frogs to generate "an interim chronic benchmark of 0.12 mg/L."

1.6 How Can Perchlorate Be Treated?

As a result of the high aqueous solubility and mobility of perchlorate, groundwater contamination is a greater concern than soil remediation at many sites. Most discussions of perchlorate remediation focus on groundwater rather than soil. Depending on the matrix and site-specific considerations, treatment via anaerobic biodegradation or thermal decomposition can destroy perchlorate. Perchlorate can be separated from water by ion exchange, adsorption onto specialized media, or membrane filtration. Table 1.1 lists the common forms of perchlorate treatment and their general applicability.

1.7 Organization of This Book

This book is organized to follow the logical sequence of identifying and solving perchlorate problems in the environment:

Chapter 2 introduces five case studies of perchlorate-contaminated sites. Subsequent chapters refer to these case studies to illustrate specific aspects of environmental problems and solutions.

Each chapter includes definitions for terms that pertain to the topic under discussion. Acronyms are listed in a section at the end of this book, as many of the acronyms are used in multiple chapters.

Readers will note that different systems of units are used throughout this book. The authors followed conventions generally used for each of the technical disciplines represented herein. For example, toxicological data are provided in metric units (e.g., dose-response value in mg/kg•day); in contrast many engineering data are recorded in English units (e.g., flow rate in gallons per minute).

Table 1.1 Applicability of Common Treatment Techniques

Form of Treatment	Technology	Applicability	
		Soil	Groundwater
Separation	Ion exchange		♦
	Standard Granular Activated Carbon		♦
	Cationic-substance coated media		♦
	Membrane filtration		♦
	Capacitive deionization/Carbon Aerogel		♦
Destruction	Bioreactors		♦
	In situ biodegradation	♦	♦
	Thermal Destruction	♦	♦
	Phytoremediation	♦	♦

CHAPTER 2

Sources, Properties, and Case Studies

This chapter provides the basis for the remainder of the book, describing the environmental sources and chemical properties of perchlorate. It also presents five case studies of sites where perchlorate contaminates groundwater, soil, or surface water. Those case studies illustrate the scientific and engineering principles described in this book.

Perchlorate is an ion containing chlorine and oxygen, with a chemical structure shown in Figure 2.1 [1] and abbreviated ClO_4^-. In general, the perchlorate ion is highly soluble and stable under environmental conditions.

Six perchlorate compounds have been manufactured in large amounts in the United States: magnesium perchlorate, potassium perchlorate, ammonium perchlorate, sodium perchlorate, lithium perchlorate, and perchloric acid [2] [3]. Manufacturers produce approximately 90% of perchlorate products, primarily ammonium perchlorate, for use in defense activities and the aerospace industry [4]. Section 2.2.2 provides further detail about the commercial production and use of perchlorate compounds. Perchlorate can also derive from natural sources or the degradation of other commercial chemical products, as described in Section 2.2.1 and 2.2.3, respectively.

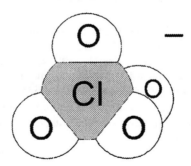

Figure 2.1 Perchlorate Structure [adapted from Urbansky and Schock, 1999] [1]

2.1 Environmental Sources

Perchlorate found in the environment can originate from natural sources, commercial use, or from the degradation of other commercial chemicals. Each of these types of sources is described below. As further discussed in Chapter 3, chemists are exploring the use of isotopic analysis to determine whether perchlorate detected in the environment originated from natural or anthropogenic sources.

2.1.1   Natural Sources of Perchlorate

Perchlorate may originate from two natural sources: soils in arid climates derived from ancient marine seabeds, and, potentially, conditions during lightning storms. Historically, Chilean nitrate fertilizer is the largest known natural source of perchlorate. Farmers have used this fertilizer in the U.S. since 1830, with its use peaking in the years 1909 to 1929 at a total of 19 million tons, or nearly one million tons per year. The use of Chilean nitrate fertilizer continues in the U.S. today, with more than 75,000 tons containing 0.01% perchlorate imported annually between 2002 and 2004. Generally, farmers in the U.S. have applied Chilean nitrate fertilizer to tobacco plants, as well as cotton and some fruit crops [5].

The major Chilean source lies in the Atacama Desert, where the fertilizer is extracted from deposits of nitrate ores or brines. The deposits contain concentrations of perchlorate ranging from 0.03 to 0.1%, or 300 to 1,000 milligrams per kilogram (mg/kg) in soil [6]. Originally a saline marine seabed, the Atacama Desert was uplifted over time to its final elevation of 2500 feet above mean sea level. The climate in the area became arid, and the lack of rainfall kept the soluble perchlorate from leaching out of the nitrate deposits. This formation has occurred in numerous other locations, and indeed, when scientists performed research into other similarly formed areas in North America, they found high concentrations of perchlorate in the soil. Researchers have detected perchlorate in soils in the following locations, all of which were formerly marine environments and are now located in areas with arid climates [6] [7].

- The Bolivian Playa crusts in the Andean high plains (with perchlorate concentrations as high as 500 mg/kg)
- Portions of 60,000 square miles in West Texas (where perchlorate ranges from 20 to 59 micrograms per liter (μg/L) in public water supply wells)
- Portions of 6,800 square miles in New Mexico (where perchlorate ranges from 0.5 to over 4 μg/L in public water supply wells)
- The Mission Valley Formation in San Diego, California (in the Otay River Valley)

Perchlorate is also found naturally in potash (potassium ore deposits) in mines near Carlsbad, New Mexico and in central Canada [6]. The potassium chloride in these deposits originated in briny sea beds, similar to those of the Chilean deposits. These deposits are dry because of their locations in deep mines. The concentrations of perchlorate in these deposits range from 25 to 2,700 mg/kg in soil.

Infrequent precipitation in arid environments where these nitrate and potash deposits formed minimizes the opportunity for perchlorate to dissolve and migrate to the groundwater. Therefore, over many thousands of years, more perchlorate would remain in these deposits than in similar deposits located in environments with higher precipitation.

In addition to occurring in ancient nitrate and potash deposits, some research has suggested that perchlorate may be created in very low concentrations during lightning storms and subsequently deposited onto surface soils. Researchers have recently simulated perchlorate production by creating an electrical discharge of chloride aerosol and by exposing the aerosol to high concentrations of ozone. The same researchers then tested their assumption by analyzing rain and snow samples from Lubbock, Texas and Cocoa Beach, Florida and found that perchlorate was present in many of the samples [8].

Providing further evidence that perchlorate may originate from natural sources, researchers at the United States Geologic Survey measured perchlorate in Pleistocene and Holocene groundwater in north-central New Mexico [9]. They found perchlorate at concentrations of 0.12 to 1.8 μg/L in groundwater samples estimated to be up to 28,000 years old based on radiocarbon dating. The researchers concluded that the perchlorate could not be anthropogenic and that it resulted from natural sources, including atmospheric deposition.

2.1.2 Anthropogenic/Commercial Sources of Perchlorate

As the case studies later in this chapter illustrate, many of the sites contaminated with perchlorate were once associated with the manufacture of perchlorate or its use in defense-related operations such as rocket manufacture or munitions use or demolition. Other environmental contamination has resulted from the use of perchlorate in fireworks, flares, or blasting agents [5]. Table 2.1 summarizes the most common commercial uses of perchlorate. It includes the available information regarding the quantity of perchlorate produced in the United States to provide the reader with a sense of how much perchlorate has been manufactured and which products predominated. As illustrated in Table 2.1, manufacturers have produced ammonium perchlorate, used in rocket fuel, in the greatest quantity.

While perchlorate's major uses are in the manufacture of munitions, explosives and fireworks, perchlorates may also be used in small amounts in other high-volume consumer products. Since 1976, over 14,000 patents have been issued for the use of various perchlorate-containing materials [20]. While many of these patents are for specialized applications in the chemical industry, complexes of transition metals such as bis-(2,2'-bispyridylamine) copper(II) perchlorate or tris(di-2-pyridylamine) iron(II) perchlorate are cited in over 90 patent applications as possible bleach activators for laundry and dishwasher detergents. Although perchlorate has been measured in the detergents used in laboratories, such as Alconox®, Liquinox®[1] and Neutrad®[2] at concentrations up to 2.5 mg/kg [21], no surveys of common laundry or dishwasher detergents have been reported. Since all of the major U.S. suppliers of these products include perchlorates as potential constituents of high-volume products, this possible source cannot be ruled out. Perchlorates are also cited as a constituent in rust removers for consumer and industrial use. Finally, perchlorate

---

[1] Alconox and Liquinox are ® registered trademarks of Alconox, Inc.

[2] Neutrad is a ® registered trademark of Decon Laboratories, Inc.

Table 2.1 Perchlorate Forms and Uses

Compound	Chemical Formula	Estimated Production in US 1951–1997[3] (million lb.)	Use
Ammonium perchlorate	NH_4ClO_4	609	Energetics booster in rocket fuel [16], used primarily by the Department of Defense and National Aeronautics and Space Administration.
Sodium perchlorate	$Na\ ClO_4$	20	Strong oxidizing agent used in the explosives and chemical industries [17].
Potassium perchlorate	$KClO_4$	22	Solid oxidant for rocket production; also used in pyrotechnics [18].
Lithium perchlorate	$LiClO_4$	Unknown	Electrolyte in voltaic cells and batteries involving lithium anodes; thin film polymers used in certain electrochemical devices may be doped with lithium perchlorate to impart conductive properties; used in synthesis of certain organic chemicals [19].
Magnesium perchlorate	$Mg(ClO_4)_2$	0.7	Drying agent for industrial gases; electrolyte for magnesium batteries; used in synthesis of certain organic chemicals [19].
Perchloric acid	$HClO_4 \cdot 2H_2O$	Unknown	Analytical reagent; when hot and concentrated, oxidizing agent and dehydrating agent [19].

has been noted in patents for certain airbag inflators. However, while the patent literature indicates this use, the extent of commercial development of this application is unclear [22].

Tracing the Origins of Perchlorate in One Drinking Water Supply
In early 2004, the MassDEP promulgated emergency regulations requiring all

---

[3] These estimates reflect production by major manufacturers and do not include all perchlorate synthesized in the United States since production reportedly began in 1908. Figures are based on estimated production rates by Kerr McGee 1951-1997 [10], estimated production rates 1959-1988 (records destroyed in fire/explosion) by American Pacific [11], and estimated production by OxyChem [12] (formerly Hooker Chemical; Columbus MS plant only from 1959-1963; production shut down in 1965), and their predecessor companies. The total estimates of production capacity and demand in the U.S. circa 1988 developed from these production rates were corroborated by The Institute for National Strategic Studies [13].

The following companies have also produced perchlorate compounds in the United States [14] [15]: Hooker Chemical (NY); Pennwalt (merged with Elf Atochem North America), GFS; Spectrum Chemicals & Laboratory Products; Mallinckrodt, Inc.; Island Pyrochemical Industries Corp.; Mil-Spec Industries Corp.; Alfa Aesar Johnson Matthey.

public water supplies to test for perchlorate contamination as the first step in determining whether to establish a drinking water standard. During the next year, this testing program provided data on 591 of 700 public water supplies in the state. Samples from 12 sources in 9 water supply systems contained perchlorate at concentrations above 1 μg/L [23]. One of the affected supplies belonged to the Town of Tewksbury. The story of how that water supply came to be contaminated shows how a relatively modest industrial use of perchlorate can have far-reaching effects.

Tewksbury draws its water supply from the Merrimack River, which has a 5,000 square mile watershed and an average mean flow rate greater than 5,000 cubic feet per second [23]. In August 2004, chemists detected perchlorate in samples of Tewksbury drinking water at concentrations between 1 and 3 μg/L. Working upstream from the Tewksbury water intake, the MassDEP collected samples from the Merrimack River and the Concord River (which discharges into the Merrimack upstream of the Tewksbury water intake). The analytical results pinpointed the perchlorate source as the 3.1 million gallon per day discharge from the wastewater treatment plant serving the town of Billerica [23]. In November 2004, the MassDEP measured perchlorate in the treatment plant influent at 0.06 to 640 μg/L, and in the effluent at 12.4 to 807 μg/L [24].

The Town of Billerica began its own investigation to determine the source of perchlorate to their wastewater treatment plant. That investigation found the apparent source. A local company, which produced surgical and medical materials, used approximately 220 gallons per month of perchloric acid in a bleaching process. Rinse water containing an average of 10 pounds per day of perchlorate discharged to the sewage system. Because the manufacturer used perchlorate in a batch process, the discharges were highly variable. The discharge complied with the facility's permit, as perchlorate was not regulated at the time [23] [24].

Once the manufacturer was identified as the source, it voluntarily shut down operations until it could treat the water. The manufacturer now treats its wastewater using ion exchange to reduce perchlorate concentrations from 2,000 mg/L to less than 0.050 mg/L [23] [24].

Releases to the environment from the manufacture and use of each of these commercial products are virtually impossible to quantify. However, some data exist with respect to outdoor, non-military uses of perchlorate-containing materials. Table 2.2 describes some of the potential for environmental releases from such sources.

2.1.3 Degradation of Other Compounds

Perchlorate can also be found as a breakdown product of sodium hypochlorite (i.e., bleach) and can be incidentally formed in corrosion control applications. These sources are described below.

The Massachusetts Department of Environmental Protection (MassDEP) sampled hypochlorite solutions used at water and wastewater treatment plants across the state

Table 2.2 Outdoor Uses of Perchlorate Compounds and Potential Release Rates [5]

Application	Estimated Annual Perchlorate Release Rate (pounds)	Notes
Chilean nitrate fertilizer	15,000	Approximately 75,000 tons fertilizer containing 0.01 weight (wt)% perchlorate used annually between 2002 and 2004.
Fireworks	Cannot be reliably estimated	Fireworks can contain up to 70 wt% potassium or ammonium perchlorate, and over 221 million pounds of fireworks were consumed in 2003 in the U.S. Environmental releases of perchlorate are difficult to predict due to thermal decomposition of perchlorate, and no data currently exist to quantify releases.
Safety flares	240,000	Preliminary research suggests that unburned and burned flares can leach 3.6 g and 1.9 mg, respectively perchlorate. Estimated 20-40 million flares used annually.
Blasting explosives	Cannot be reliably estimated	Blasting agents used in coal mining, quarrying, construction, and other uses can contain perchlorate at up to 30 wt%. The U.S. produces ~2.5 million tons of explosives annually. Environmental releases of perchlorate are difficult to predict due to thermal decomposition of perchlorate, and no data currently exist to quantify releases.
Sodium chlorate, used as defoliant/ herbicide (and potentially containing trace levels of perchlorate)	1,700	Electrochemical production of sodium chlorate can generate perchlorate as an impurity at 50-230 mg/kg chlorate. The annual consumption of sodium chlorate in the U.S. is approximately 1.2 million tons, of which approximately 94% is used in bleaching processes at pulp and paper mills. Sodium chlorate is also used as a defoliant/herbicide, particularly for cotton production in CA and AZ. Perchlorate release from trace levels in sodium chlorate defoliant use between 1991 and 2003 are estimated at 20,000 lb.

in 2005 [23] and found perchlorate concentrations ranging from 260 to 4,600 µg/L in commercial-grade hypochlorite solutions. As described by the MassDEP [23], the Town of Tewksbury, Massachusetts studied the formation of perchlorate in hypochlorite solutions over time. Fresh sodium hypochlorite solutions containing less than 0.2 µg/L were subjected to different storage conditions (i.e., capped vs. not capped, room temperature vs. chilled, dark vs. light) for 26 days and then reanalyzed. Concentrations of perchlorate reportedly increased with time to as high as 6,750 µg/L.

MassDEP also tested household bleach from a number of retailers [23]. The analytical results showed perchlorate concentrations ranging from 89 to 8,000 µg/L in the bleach samples. Bleach that had been manufactured 2.5 years before sampling occurred contained the highest concentration of perchlorate.

The mechanism hypothesized for perchlorate formation in sodium hypochlorite solutions begins with the initial degradation of hypochlorite to chlorate by a second order process [25]. Formation of perchlorate from chlorate could then occur by oxidation. Alternatively, it could occur as an intermediate by-product from an interaction of two hypochlorite degradation pathways, i.e., degradation to chlorate, and degradation to oxygen and sodium chloride [23].

Jackson et al. [26] reported another potential source of perchlorate in drinking water systems. Investigators traced the elevated concentrations of perchlorate (71 to 77 μg/L) found at a pump station in the City of Levelland Texas to an elevated storage tank. By evaluating possible mechanisms for the formation of perchlorate, researchers identified several possibilities including electrochemical generation of perchlorate such as lightning strikes or internal cathodic protection. They performed laboratory experiments that showed perchlorate could be generated in chlorinated tap water by inducing a current, using the same type of anode used in the cathodic protection system on the elevated storage tank. As there was no evidence of lightning damage, Jackson et al. [26] concluded that the internal cathodic protection system was the probable cause. They also noted that this likely was an uncommon event because of the nature of the cathodic protection system, the voltages used, and the relatively slow turnover time in this particular tank.

2.2 Physical and Chemical Properties of Perchlorate

The physical and chemical properties of perchlorate compounds determine their behavior in the environment and their potential health effects. These properties also form the basis for analytical methods and techniques for remediating environmental contamination.

Definitions
- Acid dissociation constant - an equilibrium constant that indicates the extent of dissociation of hydrogen ions from an acid
- Anion - negatively charged ion
- Cation - positively charged ion
- Complex - chemical structure consisting of a central atom or molecule weakly bonded to surrounding ions
- Monobasic acid - acid which contains only one replaceable hydrogen atom per molecule
- Orbital - position of electrons around the nucleus of an atom
- Oxidizing agent - chemical compound that accepts electrons from another compound in an oxidation-reduction reaction
- pKa - the negative log of the acid dissociation constant (a measure of acidity)
- Reducing agent- chemical compound that accepts electrons from another compound in an oxidation-reduction reaction
- Vapor pressure - pressure exerted by a vapor when it is in equilibrium with the liquid from which it is derived

2.2.1   General Properties

The properties of the most common perchlorate compounds are described below. Ammonium, sodium, potassium, lithium, and magnesium perchlorate are all solid salts at ambient temperatures, and consist of white or clear crystals. Table 2.3 summarizes the chemical formula and physical properties of the common perchlorate salts.

In contrast to the solid salts of perchlorate, perchloric acid is a colorless liquid under ambient conditions. Perchloric acid is a strong monobasic acid, with acidity greater than that of other strong acids, such as boric and nitric acids, and much greater than some common weak acids, such as carbonic acid. Table 2.4 presents a comparison of the physical properties of perchloric acid with several other common inorganic acids.

Perchlorate salts have very low vapor pressures; therefore solid perchlorate compounds cannot volatilize under ambient conditions. In addition, dissolved perchlorate anions do not tend to partition from the aqueous to the gas phase. Thus perchlorate cannot readily volatilize from water under ambient conditions. In contrast, concentrated perchloric acid could volatilize under ambient conditions, based on its moderate vapor pressure (0.9 kPa at 20° C [34]). However, if perchloric acid is diluted in water, then volatilization would be limited.

High solubility in water and the limited potential to react with other chemicals are the two physical attributes that most strongly influence the behavior of perchlorate in the environment. Each of these attributes is discussed in further detail in the following sections.

2.2.2   Solubility

When solid perchlorate salts dissolve in water they dissociate, which releases a cation and the perchlorate anion into solution. The dissociation reactions for ammonium, sodium, and potassium perchlorate are:

$$NH_4ClO_4(s) \rightarrow NH_4^+(aq) + ClO_4^-(aq) \tag{2-1}$$

$$NaClO_4(s) \rightarrow Na^+(aq) + ClO_4^-(aq) \tag{2-2}$$

$$KClO_4(s) \rightarrow K^+(aq) + ClO_4^- (aq) \tag{2-3}$$

Perchloric acid is 100% soluble in water and generally dissociates according to the following reaction:

$$HClO_4(aq) + H_2O(aq) \Leftrightarrow H_3O^+(aq) + ClO_4^-(aq) \tag{2-4}$$

The solubility values of these common perchlorate compounds range from 15 grams per liter (g/L) to 2,000 g/L, which are orders of magnitude greater than drinking water

Table 2.3 Physical Properties of Perchlorate Salts

Name (CAS Number)	Chemical Formula	Molecular Weight (g/mole)	Density (g/cm³)	Physical Appearance	Solubility in Water (mg/L)	Decomposition Temperature (°C)
Ammonium Perchlorate (7790-89-9)	NH_4ClO_4	117.49	1.95	White orthorhombic crystals [27]	200,000 at 25°C [28] 249,220 [29]	232 [30]
Sodium Perchlorate (7601-89-0)	$NaClO_4$	122.4	2.52	White orthorhombic crystals [27]	2,096,000 at 25°C [29]	492 [16]
Potassium Perchlorate (7778-74-7)	$KClO_4$	138.55	2.53	Colorless or white orthorhombic crystals [31]	15,000 at 25°C [28] 20,620 [29]	653 [17]
Lithium Perchlorate (7791-03-9)	$LiClO_4$	106.39	2.43	Small white crystals [31]	29.9% at 25°C [31]	< 250 [32] ~400, becomes rapid at 430 [31]
Magnesium Perchlorate (10034-81-8)	$Mg(ClO_4)_2$	223.21	2.21	White, hygroscopic, granular or flaky powder [31]	Very soluble in water with evolution of heat.	250 [33]

Table 2.4 Physical Properties of Perchloric Acid and Several Other Acids

Name (CAS Number)	Chemical Formula	Density (g/cm³)	Molecular Weight (g/mol)	Solubility in Water (g/100 ml)	Acidity (pKa*)
Perchloric acid (7601-90-3) [34]	$HClO_4$	1.67	100.46	Miscible	-10.00
Boric acid (10043-35-3) [27]	H_3BO_3	1.44	61.83	5.7	-9.14
Nitric acid (7697-37-2) [27]	HNO_3	1.51	63.01	Miscible	-3.37
Carbonic acid (463-79-6) [27]	H_2CO_3	1.00	62.03	Exists only in solution	6.37
* pKa = -log₁₀ Ka, or the inverse log of the acid dissociation constant. The lower the pKa value, the stronger the acid.					

exposure limits, such as California's Public Health Goal of 6 µg/L, the Massachusetts Interim Exposure Guidance Value of 1 µg/L, and the U.S. EPA's Drinking Water Equivalent Level of 24.5 µg/L. (See Chapter 6 for further discussion of these limits.)

2.2.3   Chemical Interactions

Perchlorate is one of the most powerful oxidants known to man, and yet perchlorate is also known for its lack of reactivity. On one hand, anyone who has watched a space shuttle launch has directly observed the strength of perchlorate, which is used as an oxidizing agent in the solid rocket boosters. On the other hand, chemists have long recognized the fact that the perchlorate ion is extremely stable under a wide variety of conditions.

Thermodynamic measurements indicate the perchlorate anion is a strong oxidant, however slow kinetics generally limit perchlorate reactions. Thus the reactivity of perchlorate depends strongly on temperature. Under ambient temperatures, whether dilute or concentrated, perchloric acid is not an oxidizing agent. This lack of reactivity results from the high strength of the chlorine-oxygen bonds, and the requirement that reduction proceed by removal of an oxygen atom, rather than by direct interaction of a reducing agent with the chlorine atom [35]. Earley [36] suggests that three conditions must be met before perchlorate can participate in oxidation-reduction reactions: (1) it must first form a perchlorate complex with the reducing agent, (2) there must be sufficient overlap of the electron donor and acceptor orbitals, and (3) the two orbitals must be comparable in energy. Perchlorate reduction is generally not observed except when hot, concentrated (>70%) perchloric acid is combined with a reducing agent, such as organic matter, in which case it can be explosive [37].

In addition to its resistance to oxidation and reduction reactions, the negatively charged perchlorate ion tends not to form complexes with positively charged ions. The lack of affinity for positively charged ions is due to the even charge distribution that results from the tetrahedral symmetry of the perchlorate anion. Research chemists have taken advantage of this physical property by using perchlorate mixtures to perform experiments that require unreactive, constant ionic strength solutions. Unfortunately this characteristic is a detriment from the perspective of perchlorate migration in the environment, because of the resulting low affinity for sorption to soils and consequent ability to migrate readily with groundwater flow. (Chapter 3 discusses perchlorate sorption further.)

2.2.4   Implications of Physical and Chemical Properties

Perchlorate's physical and chemical properties determine its behavior in the environment and the methods used to analyze and treat it. Table 2.5 summarizes the general properties of perchlorate and the consequences of those properties.

Table 2.5 Implications of Physical and Chemical Properties

General Property	Implications
Highly soluble; stable, noncomplexing ion	• Highly mobile in the environment once dissolved in groundwater or surface water; does not typically sorb to soils • Does not bioaccumulate • Not readily treated by precipitation or sorption • Can be treated by ion exchange
Nonvolatile	• Does not migrate as a vapor • Inhalation typically not of concern • Cannot be treated by stripping into air
Strong oxidant, but reactivity limited by kinetics under ambient conditions	• Persists in the environment • Not readily treated by chemical reduction, although biologically-mediated reactions can effectively degrade perchlorate

2.3 Case Studies: Contaminated Sites

One way to understand the environmental problems and solutions commonly associated with perchlorate is to examine sites where perchlorate contaminates soil and groundwater. The case studies below illustrate issues related to chemical analysis, fate and transport, risk assessment, and remediation.

The profile of each site begins with the setting and history of the facility. A site's setting within a community profoundly influences attitudes toward environmental concerns and, consequently, risk management decisions. Thus, background information about a site's history provides the context for understanding how the solutions to environmental problems are chosen.

Each site profile then describes the use of perchlorate and its discovery in the environment; perchlorate's fate and transport at the site; regulatory actions; and remedial actions.

Each case study was selected to illustrate a particular aspect of perchlorate in the environmental arena. Table 2.6 lists the sites and why the authors selected each for inclusion.

Table 2.6 Case Studies

Site	Features of Particular Interest
Kerr McGee and PEPCON, Henderson, Nevada	• Former sites of primary perchlorate production in the U.S. • Release of perchlorate from Kerr McGee affected the drinking water supply of 15 to 20 million people • Among the earliest remediation systems
San Gabriel Valley Superfund Site/Baldwin Park La Puente, California	• Illustrates history and consequence of refining analytical methods to lower perchlorate detection limits • Groundwater containing perchlorate is treated for use as a drinking water supply
Longhorn Army Ammunition Plant Karnack, Texas	• Remedial actions initially designed for other contaminants were retrofitted after perchlorate discovered in discharge • Extensive study of environmental effects
Indian Head Naval Surface Warfare Center Indian Head, Maryland	• Low permeability soils limited migration • Extensive testing of biological treatment methods
Massachusetts Military Reservation Cape Cod, Massachusetts	• Located over sole-source drinking water aquifer • Highly permeable soils have enabled migration • Extensive research and development of treatment methods; soil and groundwater have been remediated

2.3.1   Kerr McGee and PEPCON, Henderson, Nevada

The City of Henderson Nevada prides itself as "born in America's defense." During World War II, the Defense Plant Corporation constructed the Basic Magnesium Plant to supply the War Department with magnesium for munitions and airplane parts. Basic Management Industrial (BMI) built housing and schools for its workers nearby. In the mid-1940s, once magnesium production was no longer needed for defense, the plant closed and most of the 14,000 employees moved away. In 1949, the complex was transferred to the Colorado River Commission (CRC). The City of Henderson, incorporated on April 16, 1953, grew from the remains of the BMI complex. The CRC also conveyed a portion of the property to Western Electro Chemical Company (WECCO) and a second portion to the American Potash and Chemical Corporation (AP&CC or AMPAC). Both concerns ultimately manu-factured perchlorate compounds [10] [38].

Henderson became a center for perchlorate production, with two production plants located approximately 1.5 miles apart. These facilities, ultimately owned by Kerr McGee and Pacific Engineering and Production Co. of Nevada (PEPCON), together produced the entire supply of ammonium perchlorate for the U.S. Department of Defense and National Aeronautics and Space Administration. Ammonium perchlorate provides the oxygen needed for combustion in solid-fuel rocket motors - including the space shuttle - and numerous smaller conventional missiles [39].

Full-scale perchlorate production began in 1951 at a plant on the former BMI complex owned variously by the U.S. Navy, WECCO and AP&CC, and then ultimately by Kerr-McGee. The plant produced perchlorate until 1998. Perchlorate products included potassium perchlorate, ammonium perchlorate, sodium perchlorate, magnesium perchlorate, and sodium chlorate. The plant reportedly produced a total of 400 million pounds (200,000 tons) of perchlorate products during its operating life [10].

In 1958, PEPCON began manufacturing ammonium perchlorate at a second facility in Henderson. The operations at the PEPCON plant illustrate the manufacture of perchlorate compounds. The plant produced ammonium perchlorate in a four-step batch process [40]:

1. Electrolytic oxidation of sodium chloride to sodium chlorate;
2. Electrolytic oxidation of sodium chlorate to sodium perchlorate;
3. Reaction between sodium perchlorate and ammonium chloride, to produce ammonium perchlorate (AP); and
4. AP crystallization, filtration, drying, screening, and blending to customer specifications.

During its operating life, the plant reportedly produced approximately 234 million pounds (117,000 tons) of perchlorate products [11].

Operations at the PEPCON plant continued until May 4, 1988, when a series of fires and explosions destroyed the plant. Slag from a welding torch used to repair the building frame apparently ignited building materials and/or ammonium perchlorate residues. At the time, the facility stored an estimated 8.5 million pounds of the finished product. That material, together with the building materials, fueled fires and explosions. Two explosions registered 3.0 and 3.5 on the Richter scale at an observatory in California. The accident claimed two lives and injured over 370 people. The explosions damaged buildings within a 1.5 mile radius of the plant, and shock waves extended up to seven miles away [40] [41]. With the destruction of the PEPCON plant, the United States lost nearly half of its capacity to produce perchlorate. Due to the strategic importance of perchlorate in military applications, the Department of Defense considered this loss in capacity a threat to national security [13]. PEPCON subsequently built another perchlorate production facility at Cedar City, Utah.

Reportedly, during their operation both the Kerr McKee and PEPCON facilities discharged liquid wastes into unlined evaporation ponds. These ponds leached perchlorate into the underlying aquifer [42] [43]. The explosions at the PEPCON plant may have also contributed to perchlorate in groundwater by distributing perchlorate throughout the environment.

Three distinct contaminant plumes formed: one resulting from discharges at the Kerr McGee Plant, one from discharges at the PEPCON plant, and a third plume from the unlined waste disposal ponds. Together, these areas of contamination are known as the BMI Site. The Kerr McGee plume contains an estimated 20.4 million pounds of perchlorate in 9 billion gallons of water, and the plume from the former PEPCON plant contains an estimated 1.1 million pounds of perchlorate in 9 billion gallons of water [44].

Environmental concerns surfaced in 1997 when the Metropolitan Water District of Southern California, using a new test method with a lower detection limit, discovered perchlorate in the lower Colorado River and traced the contamination upstream to Lake Mead and the Las Vegas Wash (see Figure 2.2). These water bodies provide critically important water supplies. The Colorado River originates in the Colorado mountains and flows south for some 1,400 miles before discharging to the Gulf of California. It supplies water to seven states in the United States and to Mexico. Construction of the Hoover Dam across the Colorado River 30 miles southeast of Las Vegas created Lake Mead, a 247 square mile impoundment. The Las Vegas Wash was originally an ephemeral stream discharging to Lake Mead. With development in the Las Vegas area, however, the Las Vegas Wash now runs continuously at approximately 161 million gallons per day, primarily as a result of wastewater discharges [43] [45] [46].

The perchlorate contamination detected in these surface waters in 1997 was ultimately traced to a spring (or seep) and the discharge of contaminated groundwater from the former Kerr McGee facility. Before Kerr McGee began remedial actions, the three-mile long plume discharged an estimated 900 pounds per day of perchlorate, on average, to the Las Vegas Wash. The perchlorate releases to Lake Mead and the Lower Colorado River affected the drinking water supply of 15 to 20 million people in Arizona, southern California, southern Nevada, Tribal Nations, and Mexico [43]. Farmers also use water from the lower Colorado River to irrigate crops.

Kerr McGee began remedial actions in 1999 after entering a Consent Order with the Nevada Division of Environmental Protection. The control strategy includes containment and groundwater extraction at three strategic points [43] [45] [47]:

- At the former plant site, Kerr McGee constructed a slurry wall (1,700 ft long and 60 ft deep) and installed 22 extraction wells where perchlorate was the most concentrated (1,500,000 to 1,800,000 μg/L). The extraction wells began operation in 1999. Workers completed the slurry wall construction in October 2001. Plume capture is estimated at >99%.
- Midway between the former plant site and the Las Vegas Wash, eight extraction wells were installed in October 2002 into a narrow subsurface paleo-channel that allows for efficient capture of groundwater containing approximately 400,000 μg/L perchlorate. Plume capture is estimated at 90-95%.
- Where perchlorate-contaminated groundwater seeps into the Las Vegas Wash, seep capture began in 1999, four extraction wells were installed in 2001, and five additional wells were installed in 2003 to limit perchlorate migration into the Las Vegas Wash. Plume capture is estimated at 60-80%.

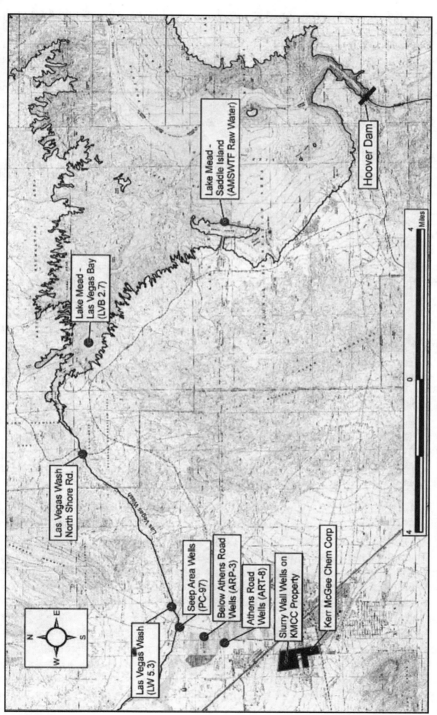

Figure 2.2 Kerr McGee Perchlorate Cleanup Project. Area plan and monitoring locations [45]

Extracted groundwater, pumped at a total rate of 1,000 gallons per minute (gpm), has been treated with ion exchange and a fluidized bed bioreactor to remove and destroy perchlorate, respectively. As of May 2005, the extraction and treatment systems removed an estimated 1,700 to 2,000 pounds per day of perchlorate, for a total of more than 1,600 tons of perchlorate throughout operation. Mass loadings to the Las Vegas Wash had decreased to estimated amounts ranging from 110 to 170 pounds per day as of late 2004. Table 2.7 summarizes monitoring results reported by U.S. EPA [45].

Table 2.7 Perchlorate Levels and Effects of Remediation, Kerr-McGee Site [45]

Monitoring Location	Perchlorate Levels 2004 (µg/L)	Apparent Effect of Remediation at the Monitoring Location
Groundwater on Kerr McGee property, upgradient of slurry wall	1,300,000-1,500,000	Little change; levels before remediation 1,500,000-1,800,000 µg/L
Groundwater on Kerr McGee property, downgradient of slurry wall	110,000-130,000	Concentration has decreased almost 90% since slurry wall installed in 2001
Groundwater downgradient of Athens Road extraction wells	90,000	Concentration has decreased approximately 80%
Groundwater at seep area	3,000	Concentration has decreased approximately 95%
Las Vegas Wash at North Shore Road	140	Concentrations have decreased from ± 700 µg/L
Lake Mead at Saddle Island	5.6 annual average *	Concentration decreased; annual average in 2000 was 13.4 µg/L
Colorado River below Hoover Dam (Willow Beach)	3.8 annual average	Average annual concentrations have declined from 6.5 µg/L in 2000

* Potentially affected by change in intake location.

As of December 31, 2004, Kerr McGee had spent $67 million on remedial actions [48]. It has been estimated based on studies of the site that remediation will require 24 (± 4) years at current removal rates [43].

American Pacific Company/PEPCON has investigated a plume of perchlorate emanating from their former facility and developed plans for remediation. At and in the vicinity of the Henderson site, perchlorate concentrations in groundwater range up to 750,000 µg/L. The plume extends approximately 300 feet through five hydro-stratigraphic units. It does not affect a water supply or discharge to surface water. PEPCON has tested *in situ* bioremediation at the site. Groundwater was pumped to the surface, augmented with ethanol and citric acid to stimulate anaerobic bacteria, and reinjected in the aquifer. Perchlorate concentrations decreased from approximately 600,000 µg/L in groundwater at the test location to less than 2 µg/L in approximately 160 days. As of this writing, the Nevada Department of

Environmental Protection is requiring that PEPCON install a remediation system at the leading edge of the plume, followed by more comprehensive remedial actions. The proposed remediation system will comprise eight extraction wells pumped at a total of 400 gpm, treatment "with a chemical that absorbs oxygen from ammonium perchlorate," and reinjection into the ground. Remediation is estimated to cost $22.4 million over 45 years [49] [50] [51].

2.3.2 San Gabriel Valley Superfund Site/Baldwin Park, La Puente, California

At La Puente, California, in the words of the city's slogan, "The Past Meets the Future." Historical use and disposal of perchlorate compounds at locations miles upgradient of the city have affected drinking water supplies for La Puente, compelling the use of state-of-the-art technologies to treat drinking water. The response to groundwater contamination problems has reflected the evolution of perchlorate analysis, risk assessment, and treatment, as shown in Table 2.8.

Table 2.8 Development of Perchlorate Clean-up Goals Throughout Project History, Baldwin Park, La Puente

Time	Project Status	Perchlorate Cleanup Goal (μg/L) [52]
Mid-1980s	EPA tests for perchlorate. Not detected using available method.	None
1997	Perchlorate detected. Drinking water wells closed.	18 *
2002	La Puente Valley County Water District treatment system operational for one year.	4 *
2004	Treatment ongoing.	6 **

* California Department of Health Services action level, which was revised over time.

** California Department of Health Services public health goal.

La Puente lies in the San Gabriel Valley in the County of Los Angeles. The underlying aquifer provides approximately 90 percent of the domestic water supply for one million residents. The surficial geology in the area comprises alluvial materials deposited by the San Gabriel River and its tributaries. Underlying sediments are typically coarse grained (e.g., sand, gravel and boulders). Northern and southern portions of the area comprise massive gravel deposits up to 500 feet or more in thickness [53].

Regulators first became concerned about groundwater quality in the Valley in 1979, when volatile organic compounds (VOCs) were detected in water samples. The U.S. EPA knew that two facilities in Azusa, California had used ammonium perchlorate and potassium perchlorate in testing solid fuel rockets and manufacturing photoflares in the 1940s. Consequently, the Agency began testing groundwater for perchlorate in the mid-1980s. However, based on the analytical methods available at the time, chemists did not detect perchlorate in groundwater samples [54].

In May 1984, the U.S. EPA listed four broad areas of contamination within the area on the National Priorities List (NPL) as San Gabriel Areas 1 through 4. The Agency subdivided the Valley into a series of "operable units" (OUs) or distinct areas of contaminated groundwater. The Baldwinville Park Operable Unit, as shown in Figure 2.3 [55], includes the La Puente Valley Water District. The U.S. EPA evaluated groundwater conditions for that operable unit between 1990 and 1993 in a Remedial Investigation and Feasibility Study.   Based on that work, the clean-up plan for the groundwater included four groundwater pump and treat systems capable of treating more than 32 million gallons per day (MGD) of contaminated groundwater.   Air stripping and/or activated carbon would remove VOCs such as trichloroethylene, perchloroethylene, and carbon tetrachloride from the water [54]. Collection of the data needed to design the pump and treat system began in 1995.

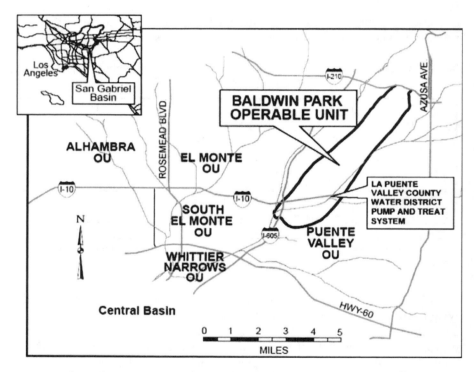

Figure 2.3 Area Plan, Baldwin Park Operable Unit of San Gabriel Valley
(Adapted from [55])

In June 1997, however, with the availability of refined analytical methods, chemists could detect perchlorate in the groundwater. Table 2.9 summarizes perchlorate monitoring data [56]. Subsequent tests also detected N-nitrosodimethylamine (NDMA) and 1,4-dioxane. The community faced two problems: detection of these contaminants forced the closure of drinking water supply wells, and the proposed clean-up methods would not remove these newly-detected contaminants [57]. The

U.S. EPA, Potentially Responsible Parties (PRPs), State agencies, and various entities with water rights reassessed remediation plans and negotiated an agreement regarding responsibility for the remedy.

Table 2.9 Perchlorate Detections in La Puente Valley County
Water District Wells [56]

Well	Number of Samples	Perchlorate Concentration Range (μg/L)	Dates of Detection
Well 02 - standby	10	50-129	1997-2003
Well 03	99	41-110	1997-2003
Well 04 - standby	6	60-159	1997-2001

The first of the four remediation systems, the La Puente Valley County Water District subproject, was completed in March 2001 at a cost of approximately $4 million. Pumps extract groundwater at approximately 2,500 gpm for treatment and subsequent use by approximately 9,000 San Gabriel Valley residents. Treatment technologies in use include air stripping, to remove VOCs, ion exchange to remove perchlorate and nitrate, and oxidation (via ultraviolet light and hydrogen peroxide) to remove NDMA and 1, 4-dioxane [55].

With respect to the larger cleanup plan for Baldwin Park, the U.S. EPA has remarked [57]:

> The cleanup plan requires pumping of approximately 22,000 gallons per minute of contaminated groundwater and treating it to remove contaminants at an estimated cost (present value) of more than $200 million, making the groundwater cleanup one of the largest and most expensive in the United States.
>
> The Baldwin Park cleanup plan combines cleanup and regional water supply goals. The negotiations needed to work out arrangements for a joint cleanup and water supply project were ultimately successful, but did not occur quickly or cheaply. The negotiations lasted more than three years and the PRP and water agency attorneys and consultants who negotiated the agreement and supported the negotiations probably cost $5 to $10 million.

2.3.3   Longhorn Army Ammunition Plant, Karnack, Texas

Activities at the Longhorn Army Ammunition Plant (LHAAP) have varied throughout the twentieth century as the United States entered and then concluded wars. Once a cotton farm, then a munitions production facility, the land now serves as a wildlife preserve.

T. J. Taylor - father of Lady Bird Johnson - sold the land to the United States government in 1941. Caddo Lake provided a ready water supply that made the site ideal for operating steam-driven machinery. Operations at this government-owned, company-operated facility began during World War II, when Monsanto started manufacturing 2, 4, 6-trinitrotoluene (TNT). The facility went on standby from 1945 to 1952. From 1952 until 1956, Universal Match Corporation produced munitions and pyrotechnic devices at the facility. Production of pyrotechnic devices such as flares and ground signals resumed again during the Vietnam War. The Thiokol Chemical Company built a facility at the plant for producing solid-fuel rocket motors for the army and operated that facility from the mid-1950s until 1971. Thiokol also began directing the production of pyrotechnic or illuminating devices in 1987. Between 1989 and 1991, under the Intermediate-Range Nuclear Force (INF) Treaty between the United States and the Soviet Union, Pershing IA and II missiles were fired and destroyed at LHAAP. Operations ceased and the facility closed in 1997. In the late 1990s, the Army decommissioned and subdivided the area. The majority of the facility is now designated for environmental preservation. The Caddo Lake Institute, funded by musician Don Henley, purchased a 25-year lease on the 1,400 acre Harrison Bayou, an environmentally sensitive area. In 2004, the Army transferred approximately 5,000 acres of the 8,493 acre facility to the US Department of Interior Fish and Wildlife Service to establish the Caddo Lake National Wildlife Refuge. The Army has retained ownership of the remainder of LHAAP land while it cleans up environmental contamination [58] [59] [60].

Until about 1984, workers washed production wastes into ponds or burned waste material in landfills. By the late 1980s, investigators had identified eleven contaminated or potentially contaminated areas. When the U.S. EPA listed the site on the NPL in 1990, the primary contaminants of concern included metals such as barium and organic compounds such as nitrobenzenes and nitrotoluenes [61]. It did not list the site as a result of perchlorate contamination. Concerns over perchlorate arose much later, after remediation of other contaminants was already underway.

Former Burning Ground No. 3 (Figure 2.4) was a special area of concern. Beginning in 1955, the facility used this 34.5-acre area to treat, store and dispose of explosives, pyrotechnics and solvent wastes by open burning, open detonation and burial. In 1963, workers constructed an unlined evaporation pond (UEP) within Burning Ground No. 3 to store wastes from the washout of rocket motor casings. Ten years later, the UEP began to receive wastes from pyrotechnic material preparation and mixing. In 1984, when investigators found contaminated groundwater, use of the UEP stopped [62].

Investigators found that subsurface soils at Former Burning Ground No. 3, comprising sands, silts and clays, are highly heterogeneous both laterally and vertically. Few strata are continuous across the site. Shallow groundwater lies at a depth from one to 23 feet below ground surface. Groundwater flows in a radial pattern away from Burning Ground No. 3, which is located on a local topographic high point. Groundwater contaminants include methylene chloride and trichloroethylene, with traces of heavy metals such as barium. Investigators observed methylene chloride and trichloroethylene as nonaqueous phase liquids. Initial site characterization work focused on these organic and inorganic contaminants, but not on perchlorate [62].

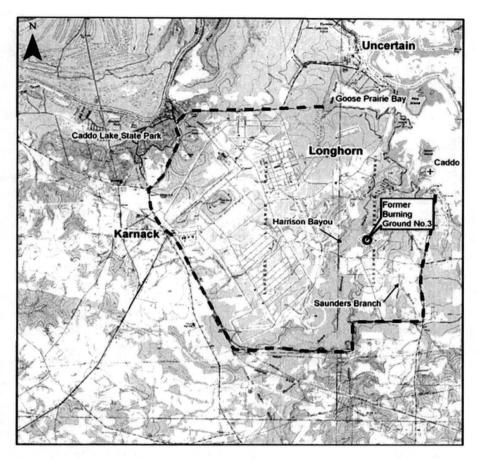

Figure 2.4 Longhorn Army Ammunition Plant (adapted from [4], Figure 2)

After determining that the concentrations of chlorinated solvents and heavy metals in the shallow groundwater and buried waste could present a significant risk, the U.S. EPA determined that groundwater should be extracted and treated. The final design included a 5,000 ft long trench and a treatment system including metals precipitation, air stripping to remove volatile organic compounds, and off-gas treatment. The discharge point for treated water varies in order to protect surface water quality. When water actively flows in Harrison Bayou, the treatment plant discharges to Harrison Bayou, then into Caddo Lake. When the bayou is dry, operators pump treated groundwater to a holding pond for storage until flow resumes in the bayou [62] [63]. The Army began operation of this groundwater extraction and treatment system in 1998.

The initial investigation and remediation efforts focused on chlorinated solvents and heavy metals. However, rockets and missiles produced at the plant used ammonium perchlorate in the solid propellant. In 1998, perchlorate was detected in the effluent from the groundwater treatment plant. This effluent discharged into Caddo Lake (Figure 2.4), which is used as a drinking water supply. In 1999, the Army collected additional samples for perchlorate analysis. Table 2.10 summarizes the results for wastewater samples collected in 1998 and other environmental samples collected in 1999 [64].

Table 2.10 Analytical Results for Perchlorate in 1999, LHAAP [64]

Location	Medium	Concentration (µg/L)
Treated wastewater	Water	10,200
		14,500
Burning Ground No. 3/Unlined Evaporation Pond	Groundwater	12,000
Goose Prairie Creek - near source	Surface water	11,000
Goose Prairie Creek - near Caddo Lake	Surface water	11
Harrison Bayou - near discharge of treated wastewater	Surface water	1,500
Harrison Bayou - near Caddo Lake	Surface water	97.3

In response to finding perchlorate in the wastewater effluent, the Army constructed a biological fluidized bed reactor to supplement the existing groundwater treatment system. The system became operational in March 2001. Treatment in the reactor reduced perchlorate concentrations from 14,000 µg/L to less than 4 µg/L [65].

A facility-wide study of perchlorate is underway as of this writing.

2.3.4   Indian Head Naval Surface Warfare Center, Indian Head, Maryland

Once the home of Algonquin Indians, Indian Head, Maryland lies on a peninsula on the Potomac River. The U.S. government established the Naval Proving Ground at Indian Head in 1890 as a naval gun testing facility. The name of the 2500-acre facility and its mission have evolved over time. Now known as the Indian Head Division, Naval Surface Warfare Center (IHDIV, NSWC), it is the National Center for Energetics. The facility provides energetics research and development, manufacturing technology, engineering, testing, manufacturing, and fleet support [66] [67].

Operations at the center generated a variety of hazardous waste. Historically, workers dumped some wastes into pits and landfills on the facility or burned wastes in open burning grounds. Of particular concern were releases of mercury from testing procedures, releases of silver from X-ray processes and manufacturing, solvent spills, and residuals from disposal areas. As a result of these releases, the U.S. EPA listed the Indian Head Naval Warfare Center on the NPL in September 1995 [68]. The Navy subsequently tested environmental samples from Indian Head for perchlorate between 1998 and 2004. Table 2.11 summarizes the results [69].

Building 1419, also known as the "Hog-out Facility," was an apparent source of perchlorate contamination. Workers there cleaned out or "hogged out" solid propel-

Table 2.11 Results of Perchlorate Analyses, 1998-2004, IHDIV, NSWC [69]

Medium	Number of Samples Tested	Number of Samples Containing Detectable Perchlorate	Maximum Concentration Detected
Drinking water	3	0	---
Groundwater	327	141	276,000 µg/L
Surface water	66	11	4 µg/L
Soil	320	121	480,000 µg/kg
Sediment	47	13	230 µg/kg

lant containing ammonium perchlorate from outdated materials such as rockets and ejection seat motors. Former waste-handling methods apparently released perchlorate to the environment [70].

Soil borings near Building 1419 showed the following surface geology: 2 to 4 feet of fill, underlain by 11 to 13 feet of clay to sandy silt. The clay and sand fraction of the silts varies. Geologists also observed discontinuous sand seams in this stratum. Below this layer, at a depth of approximately 15 feet, is a continuous layer of sand and gravel one to one and a half feet thick. A gray clay layer beneath this sand and gravel extends to at least a depth of 20 feet. Groundwater lies at a depth of 6.5 to 10.25 feet below the ground surface. Pumping tests of monitoring wells screened in the silty clay indicated a hydraulic conductivity of 0.011 to 0.044 feet per minute (ft/min). Figure 2.5 [70] shows a cross section of the geology, roughly in the direction of groundwater flow from the northwest to the southeast.

Groundwater samples from the investigation area contained perchlorate at concentrations between 1,600 and 430,000 µg/L [70]. To put these data into context, the Maryland Department of the Environment established a drinking water Advisory Level of 1 µg/L in 2002 [71].

In keeping with Indian Head's mission as a research and development facility, investigators have tested a variety of biological treatment techniques at Indian Head. Chapter 7 describes these tests further.

2.3.5   Massachusetts Military Reservation, Cape Cod, Massachusetts

The Massachusetts Military Reservation (MMR) occupies 22,000 acres, or 30 square miles, of Cape Cod. MMR includes two primary facilities: Camp Edwards, used by the U.S. Army and then the Army National Guard for training purposes, and Otis Air Force Base, operated by the U.S. Air Force and then the Air National Guard. (The Coast Guard and Veterans Administration also have facilities at MMR.) The Air National Guard and Army National Guard installations include multiple operations. This case study focuses on one area, Demolition Area 1, under the jurisdiction of the National Guard. Perchlorate and residuals from various explosives compounds contaminated groundwater and soil in that area. To understand the impact of this contamination, one must consider the site's location above a sole source drinking water aquifer.

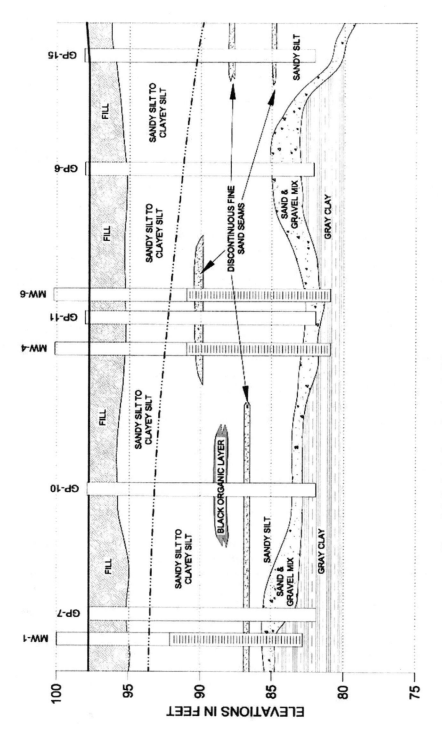

Figure 2.5 Geologic Cross Section, Building 1419 Area (from [7, Figure 11])

The needs of the nearby population for a drinking water supply have exacerbated concerns over perchlorate in groundwater at MMR (Figure 2.6). The Massachusetts National Guard began training at MMR in 1911, when the population in the surrounding towns of Bourne, Falmouth, Mashpee, and Sandwich totaled less than 10,000. When the U.S. EPA designated the aquifer beneath MMR as a Sole Source Aquifer under the Safe Drinking Water Act in 1982, the population had grown to approximately 50,000. Investigators first detected perchlorate in groundwater in 2000, when roughly 85,000 people lived in surrounding towns. By 2004, when remediation of perchlorate contamination in groundwater at MMR began, the population in surrounding towns comprised over 88,000 people who depended on groundwater for their drinking water supply. This population more than doubles during the tourist season. The population growth during the last hundred years and the resulting increase in water demand have made the preservation of the aquifer beneath MMR of critical importance, and have driven investigation and remediation efforts like those at Demolition Area 1.

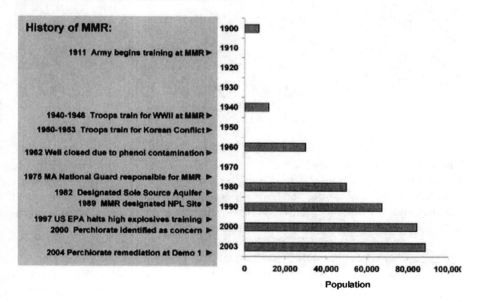

Figure 2.6 Population Growth and Activities at MMR [72] [73]

Demolition Area 1, or Demo 1, is a kettle hole, a depression in the ground surface resulting from the melting of a remnant glacial ice block. It covers approximately one acre at its base, 45 feet below the surrounding terrain. A perimeter access road surrounds the entire topographic low and associated sloping sidewalls for a total area of approximately 7.4 acres. Figure 2.7 shows an aerial view of Demo 1 [74].

In the mid-1970s, the National Guard began demolition training and explosive ordnance disposal at Demo 1. Demolition, disposal and training activities appear to have occurred primarily in the topographic low. Troops destroyed various types of

Figure 2.7 Aerial View of Demo 1 [74]

ordnance using explosive charges of C4, 2,4,6-trinitrotoluene (TNT), and det-cord. These operations continued until the late 1980s. The predominant explosive compounds used in demolition munitions are hexahydro-1,3,5-trinitro-1,3,5-triazine (also known as Royal Demolition Explosive or RDX) followed by TNT. Subsequent site investigations initially focused on these compounds and their degradation products based on known practices at Demo 1 [75].

Site investigations beginning in 1997 detected six explosive compounds (RDX, TNT, HMX, 2A-DNT, 4A-DNT, and 2,4-DNT) in groundwater and soil at Demo 1. The contaminants were all directly related to past demolition and/or disposal activities. The maximum detected concentrations of RDX and TNT in groundwater at Demo 1 were 370 μg/L and 16 μg/L, respectively [75].

In 2000, investigators found chunks of C4 and other residual munitions on the ground surface at Demo 1. Those objects, with similar materials uncovered through geophysical investigations, were removed from the site.

Investigators did not analyze groundwater samples for perchlorate until it was identified as a potential concern in early 2000. The effort to understand the occurrence and distribution of perchlorate at MMR forced the development of refined analytical methods in order to measure perchlorate at levels corresponding to regulatory clean-up goals. Analysis for perchlorate in groundwater began in April 2000 using Method 314.0 at a method detection limit (MDL) of 1.5 μg/L. Through method development by Ceimic Laboratories, the MDL was lowered to 0.85 μg/L in

December 2001 and to 0.35 μg/L in March 2002. Severn Trent Laboratories in Savannah, Georgia also analyzed groundwater samples from MMR using an MDL of 0.43 μg/L [76].

Clean-up goals for perchlorate in groundwater also evolved over time. In July 2001, the U.S. EPA established a clean-up level for perchlorate in groundwater at 1.5 μg/L [77]. The MassDEP issued an Interim Drinking Water Advice for the Town of Bourne recommending that "pregnant women, infants, children up to the age of 12, and individuals with hypothyroidism do not consume drinking water containing concentrations of perchlorate exceeding 1.0 μg/L" in 2002 [78]. Additional guidance came from the U.S. EPA Headquarters in a memorandum to Regional Administrators on January 22, 2003. That memorandum specified the use of groundwater clean-up goals between 4 and 18 μg/L perchlorate and that decision makers should consult with Headquarters when considering a value outside this range [79]. In January 2006, the U.S. EPA derived a Drinking Water Equivalent Level of 24.5 μg/L and determined that this concentration should serve as the preliminary remediation goal for perchlorate in groundwater at Superfund sites [80]. Finally, in March 2006 the MassDEP proposed a drinking water standard of 4 μg/L. This level is intended to be protective of public health considering sensitive populations such as pregnant women, nursing mothers, infants, and individuals with low levels of thyroid hormones [81].

What is the significance of these guidelines? Due to the nature of perchlorate and of the aquifer at Demo 1, perchlorate migrated readily through the aquifer, contaminating groundwater at concentrations above drinking-water limits. Groundwater contamination at Demo 1 provides a "textbook" illustration of the high mobility of perchlorate in soil and relative stability in the environment under aerobic conditions.

Munitions components other than rocket propellant that contain milligram to gram amounts of perchlorate include [76]:
- Smoke mix in 155MM Practice Projectile LITR.
- Fuse delay material in mortar rounds, signals, hand grenades, and smoke pots.
- Ignition powder in rifle smoke grenades.
- Ignition and flash composition in hand grenades and signals.
- Spotting charges in mortar practice rounds.
- Smoke charges in 22MM subcal rounds.
- Star charges, flare and illumination composition in signals.
- Incendiary composition of .50 cal Armor Piercing Incendiary and Spotter/Tracer rounds.
- Flash, whistle, and flare composition and component and photoflash charges in simulators.

The Demo 1 depression is located within the Mashpee Pitted Plain (MPP), which consists of fine- to coarse-grained sands forming a broad outwash plain. Underlying the MPP are fine-grained, glaciolacustrine sediments and basal till at the base of the unconsolidated sediments. The Demo 1 plume originates in the MPP, eventually

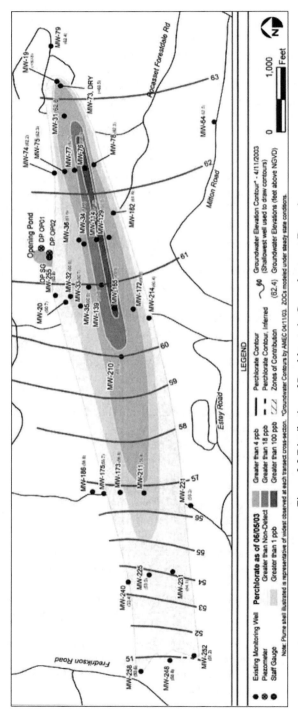

Figure 2.8 Distribution of Perchlorate in Groundwater at Demo 1

flowing into the Buzzard Bay Moraine (BBM). The BBM comprises ablation till, which is unsorted material ranging from clay to boulder size [82].

The aquifer system is unconfined (i.e., the water table is in equilibrium with atmospheric pressure and is recharged by infiltration from precipitation). The water table lies approximately 46 feet below the bottom of the Demo 1 kettle hole. The MPP sands are highly permeable, with an estimated hydraulic conductivity over 150 ft/day [82].

In the Demo 1 area of MMR, groundwater flows from the north-northeast to the south-southwest. The horizontal groundwater gradient is approximately 0.0007 ft/ft at and near the Demo 1 kettle hole. Further downgradient, in the BBM, the horizontal gradient increases to 0.0012 ft/ft on the East Side of the moraine and to 0.0023 ft/ft at the downgradient toe of the plume toward the middle of the moraine. The depth to groundwater increases by almost 14 ft, from elevation 64 ft NGVD in the source area to elevation 50.5 ft at the toe of the plume, a distance of 9,200 feet [82]. Assuming an effective porosity of 0.3 and considering the range of observed gradients and estimated conductivities, groundwater travels approximately 0.6 to 1.6 ft/day.

Figure 2.8 shows the plume of perchlorate at Demo 1 as of June 2003, before groundwater remediation began [82]. The plume originates at the Demo 1 kettle hole and extends 9,200 ft, nearly to the boundary between MMR and the nearby town of Bourne. This plume reflects the transport of perchlorate with groundwater flow, without significant sorption or lateral dispersion. As described further in Chapter 7, groundwater from this plume is being extracted, treated and discharged as a Rapid Response Action. Soil in the Demo 1 kettle hole contained perchlorate at concentrations between 1 and 110 $\mu g/kg$, with concentrations below 10 $\mu g/kg$ in about half of the soil volume, and only 10% of the soil volume containing concentrations above 50 $\mu g/kg$ [83]. This soil was excavated and thermally treated, as described in Chapter 7.

2.4 References

[1] Urbansky, E.T. and Schock, M.R., Issues in managing the risks associated with perchlorate in drinking water, *Journal of Environmental Management*, Article No. jema.1999.0274, 56, 79-95, 1999. Available online at http://www.idealibrary.com, Figure 1. (This work was completed by U.S. Government employees acting in their official capacities. As such, it is in the public domain and not subject to copyright restrictions.)

[2] U.S. Department of Health and Human Services, Public Health Service, Agency for Toxic Substances and Disease Registry, *ToxGuide for Perchlorate and Perchlorate Salts*, September 2005. Available online at www.atsdr.cdc.gov/toxpro2.html.

[3] ITRC (Interstate Technology & Regulatory Council), Perchlorate: Overview of Issues, Status, and Remedial Options, PERCHLORATE-1, Washington, D.C., Interstate Technology & Regulatory Council, Perchlorate Team, 2005. Available online at http://www.itrcweb.org, Appendix C.

[4] ITRC (Interstate Technology & Regulatory Council), Perchlorate: Overview of Issues, Status, and Remedial Options, PERCHLORATE-1, Washington, D.C.: Interstate Technology & Regulatory Council, Perchlorate Team, 2005. Available online at http://www.itrcweb.org, page 2.

[5] Cox, E., Evaluation of Alternative Causes of Wide-Spread, Low Concentration Perchlorate Impacts to Groundwater, report prepared for the U.S. Strategic Environmental Research and Development Program, Geosyntec Consultants, Boxborough, MA, May 2005.

[6] Duncan, P.B., Morrison, R.D., and Vavricka, E., Forensic Identification of Anthropogenic and Naturally Occurring Sources of Perchlorate, *Environmental Forensics*, 6, 205-215, June 2005.

[7] Jackson, W. A. et al., Perchlorate occurrence in the Texas Southern High Plains Aquifer System, *Ground Water Monitoring & Remediation,* doi: 10.1111/j.1745-6592.2005.0009, 25 (1), 137-149, 2005.

[8] Dasgupta, P.K. et al., The Origin of Naturally Occurring Perchlorate: The Role of Atmospheric Processes, *Environmental Science and Technology*, 39(6), 1569-1575, 2005.

[9] Plummer, L.N., Bohlke, J.K. and Doughten, M.W., Perchlorate in Pleistocene and Holocene Groundwater in North-Central New Mexico, *Environmental Science and Technology,* 40, 1757-1763, 2006.

[10] Smith, J.T., II, Covington & Burling, for Kerr-McGee Chemical LLC, letter to John Kemmerer, Chief, Superfund Site Cleanup Branch, Region IX USEPA, April 17, 1998.

[11] Gibson, F.D., III, Attorney at Law, for American Pacific Corporation, letter to John Kemmerer, Chief, Superfund Site Cleanup Branch, Region IX USEPA, April 14, 1998.

[12] Mack, J.A., Associate General Counsel, OxyChem, letter to Kathi Moore, Section Chief, Superfund Site Cleanup Branch, USEPA Region 9, RE: Information Request - Perchlorates, January 29, 2002.

[13] Linke, S. R., *Managing Crises in the Defense Industry: The PEPCON and AVTEX Cases*, The Institute for National Strategic Studies, National Defense University, Fort McNair, D.C., 8, July 1990.

[14] Rogers, D.E., Lt. Col. USAF, Environmental Law Directorate, Wright Patterson Air Force Base OH, memorandum for: Annie M. Jarabek, EPA National Center for Environmental Assessment (MD-52), Research Triangle Park, NC,

Subject: Perchlorate User and Producer Information, October 30, 1998.

[15] *Thomas Publishing Company*, s.v. "ThomasNet® Industrial Web Search - Perchlorate," www.thomasnet.com (accessed December 28, 2005).

[16] *American Pacific Utah Operations*, s.v. "Ammonium Perchlorate," http://www.apfc.com/utah/product1.html (accessed November 20, 2005).

[17] *American Pacific Utah Operations*, s.v. "Sodium Perchlorate," http://www.apfc.com/utah/product7.html (accessed November 20, 2005).

[18] *American Pacific Utah Operations*, s.v. "Potassium Perchlorate," http://www.apfc.com/utah/product5.html (accessed November 20, 2005).

[19] *GFS Chemicals*, s.v. "Perchlorate Compounds," http://www.gfschemicals.com/productcatalog/Perchlorate Compounds.asp (accessed December 27, 2005).

[20] *U.S. Patent Office*, s.v. "Perchlorate," http://www.uspto.gov/ (accessed August 28, 2005).

[22] ITRC (Interstate Technology & Regulatory Council), Perchlorate: Overview of Issues, Status, and Remedial Options, PERCHLORATE-1, Washington, D.C., Interstate Technology & Regulatory Council, Perchlorate Team, 2005, Available online at http://www.itrcweb.org, page 18.

[22] ATSDR, Draft Toxicological Profile for Perchlorates, U.S. Department of Health and Human Services, Public Health Service, Agency for Toxic Substances and Disease Registry, Division of Toxicology and Environmental Medicine/Toxicology Information Branch, 1600 Clifton Road NE, Mailstop F-32, Atlanta, GA, 2005. Available online at http://www.atsdr.cdc.gov/toxpro2.html#Draft, Chapter 5.

[23] Massachusetts Department of Environmental Protection, *The Occurrence and Sources of Perchlorate in Massachusetts DRAFT REPORT,* August 2005. Available on line at http://www.mass.gov/dep/cleanup/sites/percsour.pdf, 31-39.

[24] Crane, J. P., Toxin traced to neighbor, *The Boston Globe*, December 2, 2004, http://www.boston.com, (accessed February 3, 2006).

[25] Gordon, et al., Bleach Stability and Filtration, AWWA Water Quality Technology Conference, Boston, MA, November 1996, http://www.powellfab.com/products/SodiumHypo/sodium hypochlorite stability and filtration.html, Institute of Manufacturers.

[26] Jackson, A. et al., Electrochemical generation of perchlorate in municipal drinking water systems, *Journal AWWA,* 96, 103-108, 2004.

[27] Weast, R.C. (ed.), *Handbook of Chemistry and Physics*, 60th ed., CRC Press Inc., Boca Raton, Florida, 1979, p. 1757.

[28] Ashford, R.D., *Ashford's Dictionary of Industrial Chemicals*, Wavelength Publications Ltd, London, England, 1994, 758.

[29] Long, J. R., Perchlorate Safety: Reconciling Inorganic and Organic Guidelines, *GFS Chemicals Publication*, November 14, 2001, accessed at http://www.gfschemicals.com/technicallibrary/perchloricacid.pdf, January 2006.

[30] *Kerr-McGee Chemical Corp.*, s.v. "Material Safety Data Sheet - Ammonium Perchlorate, April 1, 1992," http://msds.ehs.cornell.edu/msds/siri/files/bnr/bnrmk.html (accessed March 11, 2006).

[31] Budavari, S., et al., *The Merck Index, an Encyclopedia of Chemicals, Drugs, and Biologicals, Twelfth Edition*, Merck Research Laboratories, Merck & Co., Inc. Whitehouse Station, NJ, 1996, p. 945, 970, 1231, 1317.

[32] *GFS Chemicals*, s.v. "Material Safety Data Sheet, Lithium Perchlorate, Anhydrous, GFS Chemicals, Inc., December 17, 2003," http://www.gfschemicals.com/Search/MSDS/233MSDS.PDF (accessed December 27, 2005).

[33] *GFS Chemicals*, s.v. "Material Safety Data Sheet, Magnesium Perchlorate, Anhydrous. GFS Chemicals, Inc., April 22, 2005," http://www.gfschemicals.com/Search/MSDS/54MSDS.PDF (accessed December 27, 2005).

[34] *Fischer Scientific*, s.v. "Material Safety Data Sheet for Perchloric Acid, 50-72%, updated July 2004," https://fscimage.fishersci.com/msds/18230.htm (accessed December 2005).

[35] Urbansky, E.T., Perchlorate chemistry: implications for analysis and remediation, *Bioremediation Journal*, 2, 81-95, 1998.

[36] Earley, J.E., Reduction of perchlorate by Ti (III) in ethanol, in *Perchlorate in the Environment*, Urbansky, E.T. Ed., Kluwer/Plenum, New York, 2000, chap. 29.

[37] Urbansky, E.T., Perchlorate as an environmental contaminant, *Environ. Sci. & Pollut. Res.*, 9 (3) 187-192, 2002.

[38] *City of Henderson, Nevada*, s.v. "City History," http://www.cityofhenderson.com/mayor/php/history.php (accessed December 28, 2005).

[39] Linke, S.R., Managing Crises in the Defense Industry: The PEPCON and AVTEX Cases, *The Institute for National Strategic Studies*, National Defense University, Fort McNair, D.C., p. 3-4, July 1990.

[40] Mniszewski, K.R., The PEPCON Plant Fire/Explosion: A Rare Opportunity in Fire/Explosion Investigation, *Journal of Fire Protection Engineering*, 6, No. 2, p. 63-78, 1994.

[41] Routley, J.G., *Fire and Explosions at Rocket Fuel Plant*, Henderson Nevada, May 4, 1988, Report 021 - Major Fires Investigation Project, Federal Emergency Management Agency, United States Fire Administration National Fire Data Center, 1998.

[42] *Arizona Department of Environmental Quality*, s.v. "Perchlorate Studies," http://www.azdeq.gov/function/about/perchfaq.html (accessed December 28, 2005).

[43] Hogue, C., Rocket-Fueled River, *Chemical and Engineering News*, 81, Number 33, August 18, 2003, http://pubs.acs.org/cen/coverstory/8133/8133perchlorates.html.

[44] Batista, J.R. et al., Final Report: The Fate and Transport of Perchlorate in a Contaminated Site in the Las Vegas Valley, prepared for the United States Environmental Protection Agency, Section A, March 2003.

[45] U.S. EPA, *Perchlorate Monitoring Results - Henderson, Nevada to the Lower Colorado River, June 2005 Report*, compiled by USEPA, Region 9, Waste Management Division, July 25, 2005.

[46] U.S. Department of the Interior, Bureau of Reclamation, s.v. "Hoover Dam," http://www.usbr.gov/lc/hooverdam (accessed December 27, 2005).

[47] U.S. EPA Region 9, Fact Sheet: Perchlorate in Henderson, NV - Significant Controls are Operating, p. 3, July 2004.

[48] Kerr-McGee Corporation, *2004 Annual Report*, p. 116-117.

[49] Edwards, J.G., American Pacific Corp. - Perchlorate Maker Schedules Cleanup, *Las Vegas Review-Journal*, http://www.reveiwjournal.com/lvrj_home/Aug-12-Fri-2005/news/27037167.html, August 12, 2005.

[50] American Pacific Corporation, *Quarterly Report on Form 10-Q*, submitted to the United States Securities and Exchange Commission for the quarterly period ending June 30, 2005, http://www.apfc.com/finance.html, p. 10-11.

[51] U.S. Department of Defense, *Report to the Congress: Perchlorate in the Southwestern United States, July 2005*, submitted by The Office of the Secretary of Defense, Under Secretary of Defense (Acquisition, Technology & Logistics), http://clu-in.org/download/contaminantfocus/perchlorate/epa2005_1746.pdf, p. 32-34.

[52] *California Department of Health Services*, s.v. "Perchlorate in Drinking

Water: Notification Level, Updated December 6, 2005," http://www.dhs.ca.gov/ps/ ddwem/chemicals/perchl/notificationlevel.htm (accessed December 29, 2005).

[53] U.S. EPA, Record of Decision - Baldwin Park Operable Unit, San Gabriel Valley Superfund Sites, Los Angeles County, California, March 31, 1994.

[54] U.S. EPA, EPA Updates Cleanup Plan for the Azusa-Irwindale-Baldwin Park Area. San Gabriel Valley Superfund Site/Baldwin Park Operable Unit Explanation of Significant Differences, May 1999, p. 9.

[55] U.S. EPA, Fact Sheet: Azusa/Baldwin Park Cleanup Underway - Construction Starts on Joint Cleanup and Water Supply Project, October 2002, p. 7.

[56] Dickerson, D., Executive Officer, California Regional Water Quality Board, Memorandum to: Regional Board Members, Subject: Update on Perchlorate Groundwater Pollution within the Los Angeles Region, April 28, 2003.

[57] *U.S. EPA*, s.v. "San Gabriel Valley (Area 2 - Baldwin Park), California, EPA ID# CAD980818512, San Gabriel Valley (Area 2) Overview, Updated November 2, 2004," http://www.epa.gov/superfund/sites/npl/ca.htm#San Gabriel Valley(Area 2) (Accessed December 29, 2005).

[58] *Handbook of Texas Online*, s.v. "Longhorn Army Ammunition Plant," http// www.tsha.utexas.edu/handbook/online/articles/LL/dml3.html (accessed December 30, 2005).

[59] U.S. Army, Installation Action Plan for Longhorn Army Ammunition Plant, March 2001, http://www.globalsecurity.org//military/library/report/enviro/LHAAP_IAP.pdf (accessed December 30, 2005).

[60] Korosec, T., Former Army Bomb Plant in Karnack, Texas Soon to Be Animal Refuge, *Houston Chronicle,* Houston, Texas, May 2, 2004.

[61] U.S. EPA, Federal Register Notice: 40 CFR Part 300 National Priorities List for Uncontrolled Hazardous Waste Sites, 35502-35512 Federal Register/Vol. 55, No. 169/Thursday, August 30, 1990/Rules and Regulations.

[62] U.S. EPA, EPA Superfund Record of Decision: Longhorn Army Ammunition Plant EPA ID: TX6213820529 OU 02, Karanack, TX, 05/12/1995, EPA/ROD/R06-95/092, 1995.

[63] Smith, P.N. et al., Preliminary Assessment of Perchlorate in Ecological Receptors at Longhorn Army Ammunition Plant (LHAAP), Karmack, Texas, *Ecotoxicology*, 10:305-313, 2001, https://www.denix.osd.mil/denix/Public/Library/ Water/Perchlorate/Ecosystems/lhaapeco.html.

[64] Texas Department of Health, Public Health Assessment - Longhorn Army Ammunition Plant, Karnack, Harrison County, Texas, prepared under Cooperative Agreement with the Agency for Toxic Substances and Disease Registry, July 9, 1999, http://www.atsdr.cdc.gov/HAC/PHA/longhorn (accessed on December 30, 2005).

[65] *U.S. Department of Defense, Perchlorate Work Group*, s.v. "Success Stories: Longhorn Army Ammunition Plant," http://www.dodperchlorateinfo.net/efforts/successes/army/longhorn.htm (accessed December 30, 2005).

[66] *NAVSEA*, s.v. "Warfare Centers: Indian Head Division, Naval Surface Warfare Center," http://www.ih/navy.mil/welcome.htm (accessed December 30, 2005).

[67] *The Town of Indian Head Maryland*, s.v., "Our History," http://www.townofindianheadmd.org/history.htm (accessed December 30, 2005).

[68] *U.S. EPA*, s.v. "Indian Head Naval Surface Warfare Center, Current Site Information," http://www.epa.gov/reg3hwmd/npl/MD7170024684.htm (accessed December 30, 2005).

[69] *U.S. Department of Defense, Perchlorate Work Group*, s.v. "Fact Sheet - Indian Head Naval Surface Warfare Center (NSWC)," http://www.dodperchlorateinfo.net/efforts/sites/md/sites/Indian-Head.html (accessed December 29, 2005).

[70] Cramer, R.J. et al., Field Demonstration of In Situ Perchlorate Bioremediation at Building 1419, Prepared for Naval Ordnance Safety and Security Activity, Ordnance Environmental Support Office, NAVSEA Indian Head, Surface Warfare Center Division, NOSSA-TR-2004-001, January 22, 2004.

[71] *U.S. EPA*, s.v. "State Perchlorate Advisory Levels as of 4/20/05," http://www.epa.gov/fedfac/pdf/stateadvisorylevels.pdf (accessed December 30, 2005).

[72] *Woods Hole Research Center (WHRC)*, s.v. "Land Cover and Population Changes on Cape Cod," http://www.whrc.org/capecod/land cover population/population changes.htm (accessed January 20, 2006).

[73] FitzPatrick, W.F., The Lessons of Massachusetts Military Reservation, AEPI-IFP-1001B, *Army Environmental Policy Institute*, Atlanta, Georgia, April 2001.

[74] *U.S. Army Environmental Center*, s.v. "Impact Area Groundwater Study Program - Demolition Area 1," http://groundwaterprogram.army.mil/cleanup/areas/demo1.html (accessed January 21, 2006).

[75] AMEC, Final TM 01-2, Demo 1 Groundwater Report, Massachusetts Military Reservation, Prepared for the National Guard Bureau, April 19, 2001.

[76] AMEC, Tech-Team Memorandum 02-5 Final: Site-Wide Perchlorate Characterization Report, Massachusetts Military Reservation, Prepared for National Guard Bureau, March 30, 2004.

[77] U.S. EPA, Letter (and attachments) to B. Gregson/Impact Area Groundwater Study Program, Camp Edwards from T. Borci/Office of Site Remediation and Restoration, EPA New England, July 27, 2001.

[78] Massachusetts Department of Environmental Protection, Fact Sheet: Massachusetts Interim Drinking Water Advice for Perchlorate, Office of Research and Standards, April 16, 2002.

[79] Memorandum from: Marianne Lamont Horinko, Assistant Administrator, U.S. Environmental Protection Agency, to Assistant Administrators, Regional Administrators, Subject: Status of EPA's Interim Assessment Guidance for Perchlorate, January 22, 2003.

[80] Memorandum from: Susan Parker Bodine, Assistant Administrator, U.S. EPA Office of Solid Waste and Emergency Response, to Regional Administrators. Subject: Assessment Guidance for Perchlorate, January 26, 2006.

[81] *Massachusetts Department of Environmental Protection*, s.v. "MassDEP Proposes First-In-The-Nation Drinking Water and Cleanup Standards of 2 ppb for the Chemical Perchlorate," http://mass.gov/dep/public/press/perchlorate.htm (accessed March 18, 2006).

[82] AMEC, Impact Area Groundwater Study Program, Final Feasibility Study, Demo 1 Groundwater Operable Unit, Prepared for U.S. Army Corps of Engineers and U.S. Army /National Guard Bureau, August 19, 2005, p. 3-7 and Figure 2.5.

[83] ECC, MMR Thermal Treatment Unit Completion Report Camp Edwards, Massachusetts Military Reservation, June 2005.

CHAPTER 3

How Is Perchlorate Measured?

Although perchlorate releases and its presence in groundwater were well documented from the 1950s on (as cited by the Environmental Working Group [1]), no consistent efforts were made to characterize its distribution in the environment until the late 1990s. Since that time, however, chemists have developed an array of methods for its analysis, to the point where perchlorate accounts for more single-analyte U.S. EPA methods than most, if not all, other chemicals. Perchlorate analysis methods utilize a variety of instrumental technologies, and different U.S. EPA offices have published parallel methods for their own domains. Other governmental entities, including the Department of Defense (DoD) and the Food and Drug Administration (FDA) have sponsored method development specific to their needs. University researchers and instrument manufacturers continue to pursue other options and technologies.

This rapid development and regulatory approval of analytical methods places project managers or others needing perchlorate analyses in the position of having a complicated menu of choices, each with advantages and limitations or disadvantages. Understanding the methods may be critical to obtaining data that will support subsequent uses for regulatory submittals, site characterizations, source attribution, remedial feasibility studies, risk assessment and ongoing monitoring. Each of these data usages has specific requirements, and a method appropriate for one use may not meet the needs of other requirements. For example, a less sensitive method may not provide data required to demonstrate the absence of risk, but a method that has very low detection limits may be less accurate and more costly for samples with high concentrations.

This chapter provides an overview of the progress in perchlorate analyses and discusses current methods in the context of method selection factors. Sidebars provide information on the analytical technologies themselves.

3.1 Initial Perchlorate Analysis Methods

Since perchlorate was not recognized as a contaminant of general concern in the environment until relatively recently, U.S. EPA test methods did not address perchlorate until the late 1990s. The scientific literature reported a variety of analysis methods for perchlorate as far back as the early 1900s, including gravimetric methods based on precipitation with a complex organic compound known as "nitron" [2], isotope dilution [3], potentiometric titration of perchlorate with tetraphenylarsonium chloride [4], analysis with ion-selective electrodes [5], and spectrophotometric methods [6] [7]. While these methods were relatively straightforward, most were designed for bulk analyses of materials or analyses of relatively high concentration solutions, and their limited sensitivity and selectivity were not sufficient for trace level analyses of perchlorate in environmental samples.

3.1.1 Early Ion Chromatographic Methods for Perchlorate

The development of ion chromatography (IC) in the 1960s [8] and 1970s opened the door for more specific and sensitive trace level analyses for a wide variety of anionic species, including perchlorates [9]. In 1993, the U.S. EPA published Method 300.0 [10] for ion chromatography analysis of water samples with conductivity detection for common anions and disinfection byproducts, and although perchlorate was not included in the analyte list, laboratories applied the technique with some modifications when requested to analyze environmental samples of perchlorate. These method modifications were not standardized across laboratories, and detection limits varied between approximately 400 μg/L and low mg/L levels.

Ion Chromatography

Ion chromatography (IC) is a form of liquid chromatography in which an aqueous sample is injected onto a metal column packed with a porous solid material. This material may be polymeric resin beads, silica-based particles or minerals such as alumina or aluminosilicates. These materials interact with ionic species and retard their passage through the column. As an aqueous carrier solution, referred to as the eluent or mobile phase, is forced through the column at a constant flow rate, it will wash the ions from the sample through the column. Different kinds of ions in the sample will move through the column at different rates and will be separated from other types based on their physical properties (e.g., size, polarizability, charge density). As each type of ion is washed from the column, or "eluted," the conductivity of the exiting solution will increase. Sample ions are identified based on the length of time elapsed from their injection until they exit (elute) from the column. This time is referred to as the retention time for the ion.

Discovery of elevated concentrations of perchlorate in the environment where investigators knew that releases had occurred drove the method development for perchlorate. The uncertainties associated with perchlorate's potential health effects as it migrated to groundwater spurred efforts by the California Department of Health Services (CDHS), U.S. EPA, DoD laboratories, university researchers and instrument manufacturers to develop better analytical approaches. In 1997, CDHS published an approved protocol for perchlorate analysis of waters by ion chromatography with conductivity detection and chemical suppression and initiated testing of drinking water wells and other water sources [11]. The CDHS method has a nominal reporting limit of 4 μg/L, and the resulting survey demonstrated widespread distribution of perchlorate in the state's drinking and surface waters. It was this testing program which found perchlorate contamination in La Puente and the water supply affected by the Kerr-McGee site.

Chemical Suppression

The mobile phases used for ion chromatography contain common ions and background conductivity resulting from these ions far exceeds the conductivity resulting from trace levels of perchlorate or other anions of environmental concern. Chemists use a technique known as chemical suppression to remove the mobile phase ions from the solution as it exits the chromatographic column and thereby increase the sensitivity of the conductivity detector used in the analysis. For perchlorate analysis, this technique involves the exchange of the mobile phase positive ions with hydrogen ions through either an ion exchange column or membrane. The mobile phase is often sodium hydroxide so water molecules are formed through the replacement of the sodium ion in the solution by the hydrogen ion. Chemical suppression decreases the background conductivity of the solution to the point that slight changes due to the presence of perchlorate or other trace level anions can be seen.

U.S. EPA's Office of Research and Development (ORD) followed CDHS 1999 with Method 314.0, which was similar to the California method in requiring ion chromatographic and conductivity detection for water samples but relied upon an improved chromatographic column and incorporated a requirement to monitor for levels of dissolved salts that may interfere. Method 314.0 also offered guidance for the pretreatment of samples to minimize the effects of interferences. Method 314.0 has been promulgated as the approved method for assessment monitoring of perchlorate in drinking waters under the Unregulated Contaminant Monitoring Regulation (UCMR) [12]. U.S. EPA's Office of Solid Waste and Emergency Response (OSWER) published Method 9058 in 2000 in Draft Update IVB to Test Methods for Evaluating Solid Waste, Physical/Chemical Methods (SW-846) [13] as a method for water analyses under Resource Conservation and Recovery Act (RCRA) programs. Method 9058 is also based on ion chromatography with conductivity detection, but uses a different column for separation than Method 314.0. In Figure 3.1, high conductivity from sulfate and other common anions is evident, but

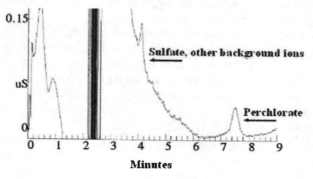

Figure 3.1 Ion Chromatograph from Method 9058 Analysis of Groundwater
(From Test Methods for Evaluating Solid Waste, Physical/Chemical Methods, Draft Update
IVB, U.S. EPA Nov. 27, 2000)

the chromatographic separation of perchlorate allows its detection by conductivity.

Both the CDHS method and EPA methods 314.0 and 9058 are essentially presumptive analyses - i.e., an increase in conductivity in the solution coming off the chromatographic column within the window of time expected for perchlorate is taken as evidence of its presence. Although none of the common anions or disinfection byproduct anions moves through the columns used in these methods at the same rate as perchlorate, there are many ionic species for which the retention times are not determined, and the presence of other ionic species can affect the rate at which perchlorate moves through the column. Polar anions such as pyrophosphate

Presumptive analysis: a chemical is identified based on properties or behavior characteristic of the chemical but that may also be characteristic of other chemicals

Definitive analysis: a chemical is identified based on properties or behavior characteristics that are unique to that chemical

($P_2O_7^{-4}$), -tripolyphosphate ($P_3O_{10}^{-5}$) and thio compounds, including aromatic sulfonates common in detergents, are potential chromatographic interferants. Sulfate ions, commonly present in drinking waters and groundwaters, can come off the column in a broad peak and obscure a perchlorate peak, resulting in a false negative.

Method 314.0 requires that the retention time for perchlorate in the sample matrix be verified by analyzing a site sample with a known amount of perchlorate added, but this is typically done on only 5% of samples, and matrix differences across a site can result in false positives or false negatives if interferences in a particular sample shift the retention time of perchlorate, come off the column at the same time as the perchlorate or obscure the perchlorate peak.

3.1.2   Improvements to Ion Chromatographic Methods

The U.S. EPA's inclusion of perchlorate in the UCMR in 1999 expanded perchlorate monitoring beyond those areas where it might be expected to be present based on manufacturing or high usage patterns. The potential for false positives or negatives became of greater concern since they could seriously undermine the overall usability of data to characterize the nationwide extent of perchlorate exposure. At the same time, increasing concerns for the possible health effects of perchlorate at trace levels drove efforts to lower the detection limits beyond the nominal CDHS and EPA method expectation of 4 μg/l. Individual laborato-

False positive: an analyte reported as present is not actually in the sample
False negative: an analyte present in the sample is reported as non-detected

ries modified IC methods using techniques such as larger injection volumes, preconcentration of perchlorate by evaporation or on short guard columns installed before the method's analytical column, sample clean-up to remove high concentrations of other anions that might obscure the perchlorate peak, or by improving the conductivity suppression [14] [15] [16] [17] [18] [19]. These techniques required validation on a case-by-case basis and performance across laboratories was not necessarily comparable. In addition, laboratories requested to analyze perchlorate in other media such as soils and foods had no guidance on appropriate techniques to extract and then analyze perchlorate from complex samples. Again, chemists developed and implemented different method modifications specific to their immediate needs.

3.2 Development of Alternative Methods

The period between 1999 and 2005 saw a high level of analytical perchlorate research, with over 100 presentations and publications on method development [19]. Instrument manufacturers improved separation columns, developed new applications for existing instruments and developed new instrumentation [14] [20]. While progress was made in lowering the sensitivity of ion chromatography with conductivity detection, other analytical techniques were also explored. Academic researchers, EPA, the FDA and DoD laboratories developed new methods to allow more definitive identification of perchlorate and its detection at lower concentrations in a wider range of sample types.

EPA researchers combined ion chromatography with mass spectrometry for the analysis of waters [21] and developed a separate method based on the formation of an ion-pair between perchlorate and a surfactant cation that could be analyzed directly by mass spectrometry [22]. Researchers at the Lawrence Livermore National Laboratory developed a method using tandem mass spectrometry (MS/MS) to characterize levels at their California facility [23]. Other techniques employing mass spectrometry were developed by instrument manufacturers [16], at Texas Tech University [18], by commercial laboratories [24] [25], and by Canada's National Research Council [26]. Numerous literature reports on improved detection sensitivity and certainty and advances in analyses for perchlorate in matrices other than water were published between 2000 and 2005. During that period, commercial laboratories incorporated modifications to Method 314.0, but there was no standardization or consistency in how the modifications were applied. Some laboratories using mass spectrometric detection after chromatographic separation referred to the approach as a modified EPA Method 8321, based on the High Pressure Liquid Chromatography (HPLC)/MS method for analyses of semivolatile organics. Other laboratories using mass spectrometry as the detector after ion chromatographic separation kept the Method 314 designation but applied a suffix or descriptor to it. Review of data for analyses conducted during this period requires some information on the actual methodology employed to understand the potential data quality issues.

Mass Spectrometry of Perchlorate

A mass spectrometer provides definitive identification of compounds eluted from ion chromatography or liquid chromatography. Mass spectrometry (MS) identifies a chemical species based on its mass and its fragmentation pattern when broken apart at the molecular or ion level. The spectrometer creates gas-phase ions and then separates the ions on the basis of their mass to charge ratio (m/z). While different anions may elute from the chromatographic column at the same time as perchlorate, they will not have the same masses as the ClO_4^- anions. A perchlorate anion will have a mass of 99 atomic mass units (amu) if the oxygen atoms are all the predominant ^{16}O isotope and the chlorine atom is the ^{35}Cl isotope. The ^{37}Cl isotope, however, occurs in nature at a ratio of approximately one ^{37}Cl to three ^{35}Cl atoms, so the number of perchlorate ions with an amu of 101 will be approximately 33% of that for perchlorate ions with an amu of 99. Perchlorate therefore produces a distinctive pattern of mass to charge ratios when subjected to mass spectrometry.

Although researchers tested a wide variety of new techniques, most programs currently rely on modifications of Method 314.0, with ion chromatography and conductivity detection, or on methods with mass spectrometry replacing conductivity as the detection tool. Method 314.0 requires relatively inexpensive instrumentation available in most commercial laboratories. Mass spectrometry is an established approach for definitive analyses of environmental samples for organics and metals. Virtually all environmental laboratories use this instrumentation and many could and did transition to mass spectrometry for perchlorate without additional capital investments. It should be noted that ion chromatography is a specialized application of liquid chromatography. With different columns and mobile phases, liquid chromatography instrumentation widely used for analyses of organics also proved applicable for perchlorate analysis. The addition of various organic chemicals to the mobile phase makes possible analyses of samples with high dissolved solids for perchlorate without the clean up steps required for ion chromatographic analyses of these samples [27].

Case Study: Massachusetts Military Reservation
    Analyses for perchlorate at the site began in early 2000, after its identification as a potential concern based on its known usage. Analyses of groundwaters were conducted by Method 314.0 initially with a method detection limit (MDL) of 1.5 $\mu g/L$ and a reporting limit (RL) of 5 $\mu g/L$. Detections between 1.5 and 5 $\mu g/L$ were reported as estimated concentrations; if no perchlorate was detected, the result was reported as <5 $\mu g/L$. During 2000 and 2001, laboratories analyzed over 2,300 groundwater samples, with perchlorate detected in approximately 6% of the samples. Ceimic in Narrangansett, RI, the support laboratory at the time, developed an improved chemical suppression approach for the ion chromatograph and was able to lower their MDL first to 0.85 $\mu g/L$ and then to 0.35 $\mu g/L$ in late 2001. The reporting limit was lowered to 1 $\mu g/L$ at that point, consistent with the MassDEP perchlorate advice level for drinking water supplies for sensitive populations. Between 2000 and 2005, over 11,000 groundwater samples from on-site monitoring wells were analyzed for perchlorate by Ceimic and STL Savannah, GA. Over half of the 2,440 detections reported are for concentrations below the detection limits of the initial analyses and would not have been detected without the laboratory's modifications to the U.S. EPA method.
    While most of the groundwater samples from the site are relatively free of interferences, there were concerns for sporadic detections of perchlorate in field blanks as well in unexpected locations. After resampling and reanalyses failed to

confirm many of these unexpected detections of perchlorate, some false positives were traced to the use of a detergent that contained perchlorate and others attributed to intermittent releases of perchlorate that had been retained on the ion chromatograph guard column.

Since perchlorate at the site is primarily attributable to surface deposition from munitions, analyses of soil quickly became of interest. The U.S. EPA had no method guidance for non-aqueous samples, but an aqueous extraction process was developed at Ceimic that reached reporting limits of approximately 10 µg/kg. With continued efforts to limit interferences, these reporting limits have been lowered to approximately 1 µg/kg for most soils.

Other media analyzed for perchlorate at the site include surface waters from ponds, wipe samples from munition fragments, influent and effluent samples from treatment processes and tissue samples from mice and vegetation. As with the soils, laboratory method modifications were developed in the absence of established methods.

While ion chromatography with conductivity detection remains the primary analysis method, HPLC/MS and more recently, HPLC/MS/MS have been used to confirm perchlorate detections in complex samples or in samples with unexpected detections.

3.3 Environmental Media - U.S. EPA Methods

During 2005, U.S. EPA ORD published Method 314.1 *Determination of Perchlorate in Drinking Water Using Inline Column Concentration/Matrix Elimination Ion Chromatography with Suppressed Conductivity Detection;* Method 331.0, *Determination of Perchlorate in Drinking Water by Liquid Chromatography Electrospray Ionization Mass Spectrometry;* and Method 332.0, *Determination of Perchlorate in Drinking Water by Ion Chromatography with Suppressed Conductivity and Electrospray Ionization Mass Spectrometry* [28].

Method 314.1 relies on ion chromatography with conductivity detection, but the perchlorate preconcentration step allows for reporting limits as low as 0.14 µg/L. The method also requires a second analysis on a different column to confirm all possible detections. This reduces but does not entirely eliminate the potential for false positives.

The mass spectral detection techniques of Methods 331.0 and 332.0 are more sensitive as well as more definitive for perchlorate. In both, the chromatographic system is interfaced directly to a mass spectrometer. The mobile phase is directed from the column into an electrospray chamber where a high voltage is applied across the liquid stream. The voltage generates droplets, which explode into smaller and smaller droplets until the perchlorate enters the gas phase as an ion. Figure 3.2 shows typical instrumentation for ion chromatography/mass spectrometry analysis.

Both Method 331.0 and 332.0 incorporate an internal standard to minimize bias from interferences or instrumental variability. A known amount of perchlorate with the ^{18}O isotope and ^{35}Cl isotope ($^{35}Cl^{18}O_4^-$, 107 atomic mass units [amu]) is added to the sample prior to analysis to be used as an internal standard. Any perchlorate in

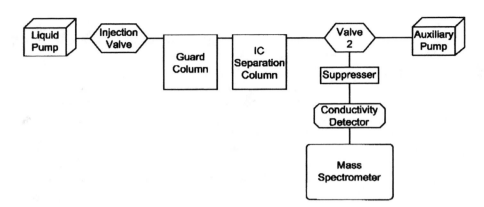

Figure 3.2 Ion Chromatography/Mass Spectrometry Instrumentation

the sample is quantified against this known amount, which provides an internal correction for any bias in the measurement due to interferences or instrument variability.

Method 332.0 is based on selected ion monitoring mass spectrometry, and this is one of two options for Method 331.0. The gas phase ions formed in the electrospray chamber are accelerated through four parallel metal rods (quadrupole) that have applied combined AC and DC potential. The voltages are adjusted to affect the trajectory of the ions passing through the rods so that at a given moment only ions of a specific mass to charge (m/z) ratio will pass through an aperture at the end of the rod assembly. Ions passing through the aperture impinge on an electron multiplier and are "counted." The mass spectrometer is programmed so only ions with m/z 99 and 101 for perchlorate and 107 for the internal standard are allowed to pass through the aperture. Figure 3.3 is an example chromatogram from the selected ion monitoring mass spectrometry analysis of a water sample, showing peaks at the same time for the characteristic ions of perchlorate and its internal standard.

Method 331.0 also provides for tandem mass spectrometry. In this instrument, once the gas phase ions pass through the first quadrupole, they enter a collision chamber where they are bombarded with argon and fragment into characteristic pieces. Perchlorate ions lose an oxygen atom in this chamber, forming "daughter" ions with m/z ratios of 83 ($^{35}ClO_3^-$) and 85 ($^{37}ClO_3^-$) while the internal standard forms an ion with the m/z ratio of 89 ($^{35}Cl^{18}O_3^-$). The fragment ions are accelerated and pass though another quadrupole with voltages adjusted so only perchlorate daughter ions will pass through the final aperture and be counted. This technique, referred to as "Multiple Reaction Monitoring" (MRM), provides highly definitive data since identification is based on the retention time of the chromatographic separation and the presence of both perchlorate ions and daughter fragments with relative abundances consistent with the natural abundance of chlorine isotopes.

Method 331.0, based on liquid chromatography followed by tandem MS analysis, can achieve reporting levels as low as 0.022 $\mu g/L$, while Method 332.0, using ion chromatography and single stage mass spectrometry, reaches reporting limits in

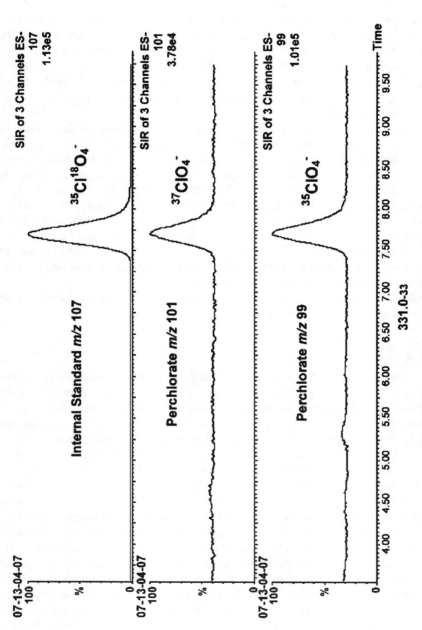

Figure 3.3 Mass Spectrometry Analysis of Perchlorate
(From U.S. EPA Method 331.0, EPA Document #: 815-R-05-007, January 2005)

water of 0.1 μg/L. Although Method 332.0 was developed for single stage mass spectrometry, liquid chromatography/tandem mass spectrometry has also been used for the analysis of aqueous samples [29].

U.S. EPA OSWER, working on a parallel path for perchlorate analyses of waters, soils and wastes under RCRA, initiated final validation studies for a proposed Method 6850, *Determination of Perchlorate Using High Performance Liquid Chromatography/Mass Spectrometry (LC/MS)* in 2005 [30]. Proposed Method 6860, which will be based on IC/MS and IC/MS/MS techniques, is also under development by OSWER. Both OSWER methods will address soil and waste samples as well as water samples.

It should be noted that although mass spectrometry can provide definitive identification of perchlorate, interferences remain a concern and may result in false negatives or elevated reporting limits. Other species that come off the separation column at the same time as perchlorate can suppress the formation of the gas phase ions in the electrospray chamber, and if an interferent has either a 99 or 101 mass, the ratio between these will be distorted and perchlorate cannot be positively identified. Sulfate is a particular concern since complete separation from perchlorate is often not achieved and hydrogen sulfate with the ^{34}S isotope ($H^{34}SO_4^-$) has the same mass as the $^{35}ClO_4^-$ anion. As for all laboratory analyses, the experience and judgment of the analyst can be critical in recognizing indicators in the data of potential interferences and in taking appropriate corrective actions.

Table 3.1 presents methods for the analyses of environmental samples with reporting limits that can be expected by each.

3.4 Biological Media

While the U.S. EPA's perchlorate method development has focused on water, soil and waste samples, the detections of the late 1990s raised public concerns for the possible effects of uptake by edible plants and into plasma and milk of both animals and humans. The detection of perchlorate in Chilean fertilizers [31] meant that perchlorate distribution may not be limited to areas with known manufacturing sources or munitions usage, and potential impacts could be widespread. The challenge for analyses of these varied media lies in the need to extract perchlorate from complex biological matrices, remove interferences and ultimately have it in a solution that then can be subjected to chromatographic separation and analysis techniques or in a form appropriate for alternative detection approaches. Perchlorate from cattle-fattening drugs had been measured by ion chromatography in cattle urine in 1993 [32], but the detection limits were not consistent with growing concerns for trace levels. Several universities and federal laboratories pursued parallel lines of research in the period following the initial discoveries of perchlorate in fertilizers, drinking waters and irrigation waters. Insects, fish, frogs, mammals and vegetation were collected within the Longhorn Army Ammunition Plant in Texas and analyzed for perchlorate [33] using accelerated extraction and cleanup techniques initially developed for beef tissue [34]. Academic researchers developed sample preparation methods [35] and adapted ion chromatographic methods for the analysis of perchlo-

Table 3.1 Methods for Analysis of Perchlorate in Environmental Samples

Year Developed	Originator	Method	Analytical Limits	Description of Method
Laboratory Methods				
1993	U.S. EPA ORD Method 300.0	Method 300.0, Determination of Inorganic Anions by Ion Chromatography	Laboratory specific	IC, conductivity detection: Perchlorate not included as analyte; method modified to accommodate perchlorate on laboratory-specific basis
1997	California Department of Health Services	Determination of Perchlorate by Ion Chromatography	MDL = 0.7 µg/L PQL = 6 µg/L	IC, conductivity detection: Applicable to water samples Subject to interference from dissolved salts Presumptive method, subject to false positives
1999	U.S. EPA ORD Method 314.0	Determination of Perchlorate in Drinking Water Using Ion Chromatography	MDL = 0.53 µg/L RL = 4 µg/L	IC, conductivity detection: Applicable to water samples Subject to interference from dissolved salts; method provides options to lower dissolved solids Presumptive method, subject to false positives

Table 3.1 Methods for Analysis of Perchlorate in Environmental Samples, continued

Year Developed	Originator	Method	Analytical Limits	Description of Method
2000	U.S. EPA OSWER EPA Method 9058	Determination of Perchlorate Using Ion Chromatography with Chemical Suppression Conductivity Detection	MDL = 0.7 µg/L	IC, conductivity detection: Essentially the same as CHDS method. Applicable to water samples. Subject to interference from dissolved salts. Presumptive method - perchlorate identified on basis of retention time
2005	U.S. EPA ORD Method 314.1	Determination of Perchlorate in Drinking Water Using Inline Column Concentration/Matrix Elimination Ion Chromatography with Suppressed Conductivity Detection	DL = 0.14 µg/L DL = 0.13 µg/L	IC: Achieves lower detection limits by concentrating perchlorate on trap; incorporates a second column analysis to confirm apparent detections from initial column analysis
2005	U.S. EPA ORD Method 331.0	Determination of Perchlorate in Drinking Water by Liquid Chromatography Electrospray Ionization Mass Spectrometry	RL = 0.056 µg/L SIM RL = 0.022 µg/L MRM	LC/MS or LC/MS/MS: Uses mass spectrometry in either selected ion monitoring (SIM) mode or multiple reaction monitoring (MRM) mode for detection and quantification; incorporates internal standard
2005	U.S. EPA ORD Method 332.0	Determination of Perchlorate in Drinking Water by Ion Chromatography with Suppressed Conductivity and Electrospray Ionization Mass Spectrometry	0.1 µg/L	IC/MS: Incorporates conductivity suppression, internal standard, single quadrupole mass spectrometer with selected ion monitoring

Table 3.1 Methods for Analysis of Perchlorate in Environmental Samples, continued

Year Developed	Originator	Method	Analytical Limits	Description of Method
2005	U.S. FDA	Rapid Determination of Perchlorate Anions in Foods by Ion Chromatography/Tandem Mass Spectrometry	1 µg/kg fruits and vegetables 0.5 µg/L bottled water 3 µg/L milk 3 µg/kg low moisture foods (oats, wheat, etc.)	IC/MS/MS: Uses 1% acetic acid to extract perchlorate from foodstuffs. Acetonitrile is added with the acetic acid to milk and low moisture foods.
In Validation, 2005	U.S. EPA OSWER Method 6850	Determination of Perchlorate Using High Performance Liquid Chromatography/Mass Spectrometry (LC/MS)		LC/MS; will include preparative methods for non-aqueous media
In Validation, 2005	U.S. OSWER Method 6860			IC/MS/MS
Field Analysis				
2004	USACE	Field Screening Method for Perchlorate in Water and Soil	1 µg/L water 0.3 µg/g soils	Colorimetric method for perchlorate in soils and waters. Uses solid phase extraction to remove perchlorate from sample. May be adapted to on-line monitor.
	Commercial suppliers	Ion Selective Electrode	100-1,000 µg/L	Limited commercial availability

rate in lettuce [36], tobacco plants and products [37], garden vegetables and grasses [38], beef cattle [39], and milk [40].

Researchers at the FDA laboratories [41] applied ion chromatography with tandem mass spectrometry to the analysis of various foodstuffs, and this work was the basis for the FDA method *Rapid Determination of Perchlorate Anions in Foods by Ion Chromatography/Tandem Mass Spectrometry* published in 2005 [42]. While the final instrumental analysis of the FDA method is comparable to EPA Method 332.0 and the proposed Method 6860, specific sample preparation techniques are provided for waters, milk, vegetables, baby food and dry foods such as oats and wheat. Research continues on preparative methods to minimize interferences from vegetation without extensive clean up procedures [43] and on improvements to ion chromatographic analysis methods that would not require costly mass spectrometric techniques. Sample cleanup by carbon solid phase extraction proved effective in removing interferences from various vegetables and subsequent analysis by ion chromatographic/conductivity resulted in data comparable to those from tandem mass spectrometric measurements [44].

As part of continuing research into perchlorate exposure and health effects, methods have also been optimized to analyze perchlorate in human urine. The sensitivity of electrospray with tandem mass spectrometry (0.025 ng/mL or μg/L) has proven sufficient to detect perchlorate in all human urine samples evaluated [45].

3.5 Field Methods

The U.S. EPA and FDA methods are essentially laboratory methods, requiring complex instrumentation with fixed plumbing and power. While they could be supported in properly equipped mobile laboratories, these methods are not ideal for rapid field analyses or site characterizations. Field methods would be of particular interest to DoD facilities where training exercises have left perchlorate residues at possibly high concentrations over large areas. The U.S. Army Corps of Engineers (USACE) has supported the development of a rapid and inexpensive colorimetric field method [46] with detection limits of 1 mg/L for waters and 0.3 milligrams per gram (mg/g) for soils. While these detection limits are not sufficiently low to completely characterize a site, they do offer the chance to identify "hot spots" and focus investigations.

Options for field analyses with somewhat lower sensitivities include ion selective electrodes and emerging technologies based on attenuated total reflectance - Fourier transform infrared spectroscopy (ATR-FTIR) [47] [48]. ATR-FTIR spectroscopy uses a silicon crystal coated with a material that will selectively absorb perchlorate. The infrared beam is guided inside the silicon crystal and light absorption within the field is measured. The Fourier transformation of the measured intensities produces an absorption spectrum of the sample. ATR-FTIR probes have been developed that can detect down to 3 μg/L of perchlorate with a 30 minute contact time. Development of these detectors suitable for use in cone penetrometers would allow *in situ* measurement.

Although ion selective electrodes with detection limits below 100 μg/L have been developed [49] [50], those currently available on a commercial basis are limited to a detection limit of 200 μg/L or higher.

3.6 Forensic Analyses

Perchlorate can originate from both natural and anthropogenic sources, as described in Chapter 2. Chemists have developed forensic analyses to distinguish whether perchlorate in an environmental sample may have originated from natural material, such as Chilean nitrate fertilizer, or an anthropogenic material. Forensic analyses can also indicate the progress of perchlorate biodegradation.

Stable isotope analyses of perchlorate indicated that differences may exist in the ratio of the chlorine atoms, ^{35}Cl and ^{37}Cl that might be useful for forensic applications [51]. A side benefit of the development of a new resin technology treatment process for perchlorate by the Oak Ridge National Laboratory (ORNL) in 2004 was the discovery that naturally occurring perchlorate has a unique oxygen isotope signature [52]. The ORNL treatment process breaks down perchlorate, forming chloride. During isotopic analysis of the chloride released from natural and manufactured sources, researchers noted consistent differences in the $^{37}Cl/^{35}Cl$ and $^{18}O/^{17}O/^{16}O$ isotope relationships that result from the formation mechanisms [53] [54] [55]. This finding has since formed the basis for source attribution of perchlorate in several instances. A variety of different sources, including manufactured perchlorate salts used for explosives, natural perchlorate-bearing salt deposits, and salt-derived fertilizers, were analyzed in order to make perchlorate source determinations for waters from contaminated military and industrial sites as well as for waters with no known local sources of synthetic material [56]. NASA is currently planning to use this technique to determine the extent of groundwater impact from past activities at the Jet Propulsion Laboratory in California [57].

Laboratory studies have also found that chlorine isotope analysis is a sensitive technique to document perchlorate biodegradation and potentially to distinguish this process from other non-biological mechanisms that reduce perchlorate levels during *in situ* remediation efforts. The perchlorate-reducing bacterium, *Azospira,* preferentially reduces perchlorate anions with the ^{35}Cl isotope. The chloride produced from bioremediation has a significantly higher $^{35}Cl{:}^{37}Cl$ ratio than chloride from background salts, and shifts in the relative abundance of the two isotopes in water are indicative of microbial degradation as a removal pathway [58]. Stable isotope analysis has been used to monitor bioremediation and to differentiate between reductions due to biodegradation and non-degradative attenuation [59].

3.7 Ongoing Method Development

Although U.S. EPA Methods incorporating improved techniques for ion chromatography and mass spectrometry options have been published (314.1, 331.0 and 332.0) or are near completion (6850, 6860), researchers continue to explore method enhancements or alternative options for perchlorate analysis. Ion chromatography followed by gas-phase ion-association mass spectrometry has recently been reported for biological samples [60]. This approach involves the formation of an adduct between the perchlorate ion and a large di-positive organic ion that then gives a stronger signal in the mass spectrometer than perchlorate alone.

Law enforcement laboratories currently use capillary electrophoresis for forensic analyses of perchlorate related to explosions [61]. While OSWER has included Method 6500, which uses this analytical method, in Draft Update IVA to SW-846 [62], perchlorate is not among the analytes addressed by the method. Detection limits for capillary electrophoresis are generally higher than those achieved by ion chromatography, but the method has proven appropriate for selected circumstances.

Capillary electrophoresis (CE) is a separation technique that uses narrow-bore fused-silica capillaries to separate a complex array of large and small molecules. High electric field strengths are used to separate molecules based on differences in charge, size and hydrophobicity. Samples are introduced into the capillary immersing the end of the tube into a sample vial and applying pressure, vacuum or voltage. Detection may be by UV absorbance or other techniques.

3.8 Developing a Strategy for Perchlorate Analyses

Given the wide range of analytical options, those selecting methods for perchlorate must consider several questions, including:

- What are the media of concern?
- What, if any, are the regulatory requirements?
- What are the program data quality objectives?
- How critical is it to have a fully definitive method?
- What are the budget constraints?

Each of these is discussed below.

3.8.1   What Are the Media of Concern?

All of the analytical methods currently in commercial use require that the perchlorate be in an aqueous phase for the instrumental analysis. Methods designed for water analysis do not address the extraction or preparation requirements for other matrices. Individual laboratories have developed extraction approaches for soils and wastes, but until the revision to EPA Method 9058 or Methods 6850 and 6860 are released, there is no prescribed protocol for soils, sludges or waste and comparability between data sets for these matrices cannot be assumed without verifying that equivalent procedures were used.

While specific procedures for biological media and food matrices have been developed as discussed in Section 3.4, the federal and academic research laboratories involved in these efforts are not readily available for environmental programs for industry. These complex matrices require more extensive sample preparation and typically are analyzed with mass spectrometric detection to reduce concerns for false positives.

3.8.2   What, if Any, Are the Regulatory Requirements?

The U.S. EPA required Method 314.0 for monitoring during the 2001-2005 cycle of the Unregulated Contaminant Monitoring Rule (UMCR) and has typically required this method to demonstrate compliance with National Pollution Discharge Elimination System (NPDES) permits. UCMR2, the proposed second cycle of monitoring for unregulated contaminants, planned for 2007-2010 [63] includes Methods 314.1, 331.0 and 332.0 as applicable perchlorate methods. The proposed program will require laboratories to demonstrate minimum reporting limits of 0.5 μg/L and will require a definitive confirmation by second column ion chromatographic analysis or mass spectrometry of any detections.

Laboratories must be formally approved by U.S. EPA to analyze perchlorate in support of UCMR monitoring requirements. For other parameters, state or primacy agency certification is sufficient, but perchlorate is an exception, requiring laboratories to pass U.S. EPA's performance testing program. Laboratories approved for perchlorate monitoring are listed at: http://www.epa.gov/safewater/standard/ucmr/aprvlabs.html.

The Department of Defense Environmental Data Quality Work Group (EDQW) has issued guidance for perchlorate characterization at DoD installations [64]. This guidance indicates that Method 314.0 alone will not be sufficient; any detections by ion chromatography with conductivity detection should be confirmed by, for example, mass spectral analysis. The Air Force Center for Environmental Excellence (AFCEE) Quality Assurance Project Plan (QAPP) [65] allows for either Method 314.0 or Modified Method 8321A, using HPLC with electrospray mass spectrometry for investigations at U.S. Air Force installations.

SW-846 serves as U.S. EPA's official compendium of analytical and sampling methods for use in complying with RCRA regulations. Although Method 9058 for water analyses was published as part of Draft Update IVB to SW-846 [66], this update has not been formally incorporated into SW-846. OSW plans to update Method 9058 and to include directions for the analyses of soils and sediments [67], but the U.S. EPA currently offers no method direction for sample preparation for media other than waters. Method 6850 should be released as a Draft Method in 2006, providing a LC/MS/MS method for RCRA generally comparable to Method 331.0 for waters and including sample preparative methods for soils and wastes.

3.8.3   What Are the Program Data Quality Objectives?

Data quality objectives (DQOs) for environmental sampling and analyses programs typically include goals for representativeness, sensitivity, precision, accuracy, comparability, and completeness, each of which is discussed further below. DQOs are established based on the level of each that will be required to make appropriate decisions with confidence. An initial screening program to find "hot spots," for example, may have significantly different DQOs than a remedial verification program conducted at a later time at the site. Defining the project requirements up front can often save considerable time and effort as well as cost.

Representativeness: Representativeness expresses the degree to which data accurately and precisely represent an environmental condition, characteristic of a population or parameter variations. Representativeness for most environmental programs depends on the sampling program design, requiring the collection of samples spatially and temporally representative of site conditions. In order to achieve acceptable representativeness, sample results also must not be affected by conditions that would lead to false positives or false negatives. Perchlorate sampling and analyses must be conducted appropriately for the media of concern, avoiding the use of detergents or other materials that may contain perchlorate, reviewing data for evidence of interferences, and properly confirming detections.

Sensitivity: Table 3.1 includes general expectations for reporting limits for the various U.S. EPA methods. The most sensitive method may be necessary to demonstrate complete removal of perchlorate in a treatment system or to accurately map plume movement. Using a highly sensitive method may be counter-productive, however, at sites where concentrations are relatively high. The linear range of any instrumental analysis is limited, and if significant dilution of a sample is required to bring it into the method range, the potential for bias from dilution errors increases.

Detection Limits

The method detection limit (MDL) for an environmental analysis is not a bright line, but a statistical construct based on the variability in results of replicate analyses of standard solutions at low concentrations. The U.S. EPA designed procedures to determine the MDL to establish the concentration above which an apparent detection has a 99% probability of being a true positive. This does not ensure that false negatives - the failure to detect the analyte when present above that level - will not occur. Individual sample characteristics, such as other chemicals present, or variability in the instrument performance may limit the detection of an analyte in a particular sample at the calculated MDL. This approach also does not mean that a peak will not be noted if the analyte is present at a concentration below the MDL, but it does mean that the confidence in that peak as a positive detection falls below U.S. EPA's criterion for reporting.

Non-detected results for environmental samples are generally reported at practical quantitation limits (PQLs) or Reporting Limits (RLs). These are typically set at a sensitivity level 3 to 5 times the MDL to provide a margin of variability. For most methods, the lower limit of the instrument calibration must correspond to the PQL/RL to ensure sensitivity for each set of sample analyses.

Precision and accuracy: Precision is a measure of how closely replicate measurements agree, while accuracy is a measure of how close the measured value is to the "true" value. A double blind study recently conducted in Massachusetts approved laboratories [68] required analyses of a series of aqueous samples ranging in concentration from <0.4 to slightly greater than 1 μg/L. Chemists performed the analyses in accordance with EPA Method 314.0 as modified by the individual laboratories to reach lower reporting limits. Overall, the study found a low bias, with measured values averaging at 83% of the true values. A subsequent study by the same author [68] included laboratories using Method 314.1, IC/MS, LC/MS/MS and a laboratory-specific dual column method. The results demonstrated that all of the methods are reliable to 0.5 μg/L for waters with low dissolved solids, with accuracies of 80 to 120% in this range. Method 314.0, however, was not as accurate for samples with high dissolved solids as were the other methods.

Commercially available performance evaluation samples are available for perchlorate in water, soil, sludge and vegetation. These are samples with certified concentrations prepared by independent vendors. They may be purchased and submitted to a laboratory to demonstrate the accuracy of their analyses as part of the quality assurance program for an investigation.

Comparability: Comparability is a measure of whether data from one investigation can reliably be compared to data from other investigations at the same or other sites for the same parameters. Historical data collected using Method 314.0 may not always be comparable to data collected using newer analytical methods with lower sensitivities or more definitive confirmation of detections. This is likely an issue for many of the longer-term investigative programs, and resampling of waters where perchlorate was not detected may be necessary to demonstrate actual levels.

Completeness: Completeness is defined as the amount of data that is valid and appropriate for the program usage in proportion to the amount of data collected. Completeness for perchlorate analyses will depend on whether the methods selected provide the sensitivity and selectivity required for the program. Data with elevated reporting limits may be analytically sound but not contribute to program completeness if greater sensitivity would be required to make necessary decisions.

3.8.4 How Critical Is It to Have a Fully Definitive Method?

Recall that analytical methods may be presumptive or definitive. Presumptive analyses identify a chemical based on properties or behavior that are characteristic of that chemical, but may also be characteristic of other compounds. Definitive analyses provide more certain identification of the analyte.

Methods 314.0 and 9058, with conductivity detection, are considered presumptive. False positives from unknown co-eluting interferences cannot always be eliminated. In addition, guard columns used for in-line sample clean up need close monitoring since intermittent breakthrough of interferences from previous samples can also result in false positives. These methods are, however, appropriate as well as cost-effective for many program. If, for example, there is evidence of releases of perchlorate at a site, or if the analyses are part of an ongoing program monitoring levels of perchlorate in groundwater where historical data are reliable, or if influent/effluent levels from a treatment system need to be monitored, the need for definitive analyses may be periodic or limited to instances of unexpected detections.

The second-column confirmation requirement of Method 314.1 provides an additional level of confidence to detections by ion chromatography. False positives, however, remain possible, especially in complex samples with multiple potential interferents. The methods employing mass spectrometry, 331.0, 332.0 and 6850, should be considered when a site is unknown, the transport pathways are not well understood or multiple interferences are possible. This does not necessarily mean, however, that a program should lock into using these more expensive methods for all analyses. Non-detects from conductivity methods are generally reliable unless the sample has high background conductivity or high sulfate, but unexpected detections warrant closer attention. A sample perchlorate analysis decision tree, shown as Figure 3.4, has been applied successfully at a large site with commingled plumes of perchlorate and organic contaminants [69]. Approximately 58% of samples from this site selected for confirmatory analysis were demonstrated by mass spectral analysis to be false positives from Method 314.0 analysis. A second set of samples from the same site, non-detect for perchlorate but coming from locations where perchlorate had previously been reported, were also analyzed by mass spectrometry. No false negatives for Method 314.0 were found.

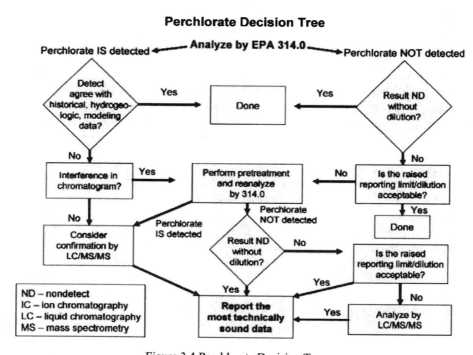

Figure 3.4 Perchlorate Decision Tree
(From Wessling, E. Chemical Quality Remains Critical. Presented at the 2005 DoD Environmental Monitoring and Data Quality Workshop. April, 2005)

While the cited program demonstrates the potential for false positives in complex matrices and the importance of confirmatory analyses, the potential for false positives is site and matrix specific. In contrast to results from that program, no false positives or trend of bias from ion chromatography results were found in the confirmatory LC/MS analyses of 113 critical groundwater samples from a site where rocket motors were constructed and tested but no other known agricultural or manufacturing sources of contaminants were present [70].

3.8.5   What Are the Budget Constraints?

Although methods relying on mass spectrometry for identification have a much higher degree of certainty for identification of perchlorate, their cost is significantly greater than methods with conductivity detection. The instrumentation required for Methods 331.0, 332.0 and 6850 is expensive - on the order of 2 to 5 times that of instrumentation for Methods 314.0 and 314.1. Relatively few commercial laboratories offer IC/MS or LC/MS/MS for perchlorate, so competition is not strong. Both factors result in per sample costs that are 2 to 4 times higher for IC/MS and LC/MS/MS than for IC with conductivity detections.

3.8.6   Sampling Considerations

Perchlorate is chemically stable under environmental conditions and not subject to aerobic biodegradation. The potential for anaerobic biodegradation was not generally recognized until the early 2000s and Methods 314.0 and 9058 only required that samples be chilled to $4° \pm 2°C$ for storage. The more recent methods impose sampling and storage requirements to prevent losses from biological action; samples should be filtered in the field, placed in bottles with 1/3 of the volume as headspace, shipped at $10°C$ or below and stored at $6°C$ or below without freezing.

3.9 Summary

As a result of the rapid method development of the past 5 years, there are currently several options commercially available for the analysis of perchlorate in environmental and biological media at trace levels. Site characterizations and exposure assessments are now possible using actual measured values for perchlorate in a wide variety of samples. These methods will be critical for understanding the sources, distribution and impact of perchlorate in the environment.

The available methods cover a range of selectivity and sensitivity as well as cost. Data usage requirements should be clearly understood before a method is selected or an overall strategy designed.

3.10 References

[1] Environmental Working Group (EWG), Rocket Science: Perchlorate and the toxic legacy of the cold war, July 2001.

[2] Cope, W.C. and Barab, J., "Nitron" as a gravimetric reagent for the analysis of substances used in explosives, *J. American Chemical Society*, 39, 504, 1917.

[3] Johannesson, J.K., Determination of perchlorate by isotopic dilution with potassium perchlorate - Chlorine 36, *Analytical Chemistry*, 34, 1111, 1964.

[4] Baczuk, R. and Dubois, R., Potentiometric titration of perchlorate with tetraphenylarsonium chloride and a perchlorate ion specific electrode, *Analytical Chemistry*, 40, 685, 1968.

[5] Rohn, T. and Guilbault, G., New methods for the preparation of perchlorate ion-selective electrodes, *Analytical Chemistry*, 4, 590, 1974.

[6] Fritz, J.S., Abbink, J.E., and Campbell, P.A., Spectrophotometric Determination of Perchlorate, *Analytical Chemistry*, 36, 2123, 1964.

[7] Trautwein, N. and Guyon, J., Spectrophotometric determination of the perchlorate ion. *Analytical Chemistry*, 40(3), 639-641, 1968.

[8] Walton, H.F., Ion exchange and liquid chromatography, *Analytical Chemistry*, 48(5), 52r-66r, 1976.

[9] Deguchi, T., Hissanaga, A., and Nagai, H., Chromatographic behaviour of inorganic anions on a sephadex G-15 column, *Journal of Chromatography A*, 133, 173, 197.

[10] U.S. EPA, *Methods for the Determination of Inorganic Substances in Environmental Samples*, Office of Research and Development, EPA/600/R-93/100, 1993.

[11] State of California Sanitation & Radiation, Department of Health Services *Determination of Perchlorate June 3, 1997, by Ion Chromatography*, Rev. No. 0 CLO4METH.

[12] U.S. EPA Office of Research and Development, Method 314.0: Determination of Perchlorate in Drinking Water Using Ion Chromatography, www.epa.gov/OGWDW/methods/met314.pdf, discussion of Method 314.0 on page 4, November 1999.

[13] U.S. EPA, *Test Methods for Evaluating Solid Waste, Physical/Chemical Methods (SW-846)*, Office of Solid Waste and Emergency Response, 1999.

[14] Jackson, P.E., Laikhtman, M., and Rohrer, J.S., Determination of Trace Level Perchlorate in Drinking Water and Ground Water by Ion Chromatography, *Journal of Chromatography A*, 850, 131, 1999.

[15] Jackson, P.E. et al., Improved method for the determination of trace perchlorate in ground and drinking waters by ion chromatography, *Journal of Chromatography,* 888, 151, 2000.

[16] Joyce, R.J. et al., Advances in the determination of perchlorate in drinking water and ground water using IC and IC/MS methods, presented at PITTCON 2003, Pittsburgh, PA, Presentation 2060-3.

[17] SERDP, *In situ* bioremediation of perchlorate, SERDP Project CU-2263, May 21, 2002.

[18] Tian, K., Dasgupta, P., and Anderson, T., Determination of trace perchlorate in high-salinity water samples by ion chromatography with on-line preconcentration and preelution, *Analytical Chemistry,* 75, 701, 2003.

[19] U.S. EPA, Measurement and Monitoring Technologies for the 21st Century, April 2005. Available through http://clu-in.org/download/misc/21M2flier.pdf.

[20] Jackson, P.E. and Chassaniol, K., Advances in the determination of inorganic ions in potable waters by ion chromatography, *Journal of Environmental Monitoring,* 4, 10, 2002.

[21] Urbansky, E. et al., Survey of bottled waters for perchlorate by electrospray ionization mass spectrometry (ESI-MS) and ion chromatography (IC), *J. Science of Food and Agriculture,* 80, 1978, 2000.

[22] Magnuson, M., Urbansky, E., and Kelty, K., Determination of perchlorate at trace levels in drinking water by ion-pair extraction with electrospray ionization mass spectrometry, *Analytical Chemistry,* 72, 25, 2000.

[23] Koester, C.J., Beller, H.R., and Halden, R.U., Analysis of perchlorate in groundwater by electrospray ionization mass spectrometry/mass spectrometry, *Environmental Science & Technology,* 34, 1862, 2000.

[24] Penfold, L. and Dymerski, M., Novel LC/MS/MS and IC/MS/MS methods for definitive identification of perchlorate in environmental samples, presented at PITTCON 2004, Chicago, Illinois, March 7-12, 2004.

[25] Winkler, P., Minteer, M., and Willey, J., Analysis of perchlorate in water and soil by electrospray LC/MS/MS, *Analytical Chemistry,* 76, 469, 2004.

[26] Ells, B. et al., Trace level determination of perchlorate in water matrices and human urine using ESI-FAIMS-MS, *J. Environmental Monitoring,* 2, 393, 2000.

[27] Krol, J., The determination of perchlorate anion in high total dissolved solids

water using LC/MS/MS, presented at NEMC 2004: The 20th Annual National Environmental Monitoring Conference, Washington, D.C., July 19-23, 2004, Book of Abstracts, 3.

[28] U.S. EPA Office of Research and Development, *Method 332.0 Determination of Perchlorate in Drinking Water by Ion Chromatography with Suppressed Conductivity and Electrospray Ionization Mass Spectrometry,* EPA/600/R-05/049, March 2005.

[28a] U.S. EPA Office of Research and Development, *Method 332.0 Determination of Perchlorate in Drinking Water by Liquid Chromatography and Electrospray Ionization Mass Spectrometry,* EPA/815/R-05/007, January 2005.

[29] Snyder, S., Vanderford, B., and Rexing, D., Trace analyses of bromate, chlorate, iodate and perchlorate in natural and bottled waters, *Environmental Science & Technology,* 39, 4856, 2005.

[30] Di Rienzo, R.P. et al., Analysis of perchlorate in drinking water, groundwater, saline water, soil, and biota by LC/MS, presented at NEMC 2004: The 20th Annual National Environmental Monitoring Conference, Washington, D.C., July 19-23, 2004, 31.

[31] Susarla, S. et al., Perchlorate identification in fertilizers, *Environmental Science & Technology,* 33, 3469, 1999.

[32] Batjoens, P., DeBrabander, H.F., and Kindt, L.T. Ion chromatographic determination of perchlorate in cattle urine, *Analytica Chemica Acta,* 275, 335, 1993.

[33] Smith, P.N. et al., Preliminary assessment of perchlorate in ecological receptors at the Longhorn Army Amunition Plant (LHAAP), Karnack, Texas, *Ecotoxicology,* 10, 305, 2001.

[34] Anderson, T.A. and Wu, T.H., Extraction, cleanup, and analysis of the perchlorate anion in tissue samples, *Bulletin of Environmental Contamination and Toxicology,* 68, 684, 2002.

[35] Canas, J., Tian, K., and Anderson, T., Application of an ion chromatography method for perchlorate determination in difficult matrices, presented at the 24th Annual Meeting of the Society of Environmental Toxicology and Chemistry, Austin, TX, November 9-13, 2003, 247.

[36] Environmental Working Group (EWG), Suspect salads toxic rocket fuel found in samples of winter lettuce, www.ewg.org/reports/suspectsalads/release 20030428.php, April 28, 2003.

[37] Ellington, J.J. et al., Determination of perchlorate in tobacco plants and tobacco products, *Environmental Science & Technology*, 35, 3213, 2001.

[38] TIEHH 2003, Perchlorate Sampling of Farm, Cherokee County, KS, prepared for Kansas Department of Health and Environment, The Institute of Environmental and Human Health, Texas Tech University, September 2003.

[39] Cheng, Q. et al., A study of perchlorate exposure and absorption in beef cattle, *J. Agricultural Food Chemistry*, 52, 3456, 2004.

[40] Kirk, A.B. et al., Perchlorate in milk, *Environmental Science & Technology*, 37, 4979, 2003.

[41] Krynitsky, A.J., Niemann, R.A., and Nortrup, D.A., Determination of perchlorate anion in foods by ion chromatography-tandem mass spectrometry, *Analytical Chemistry*, 76, 5518, 2004.

[42] U.S. Food and Drug Administration, *Rapid Determination of Perchlorate Anions in Foods by Ion Chromatography/Tandem Mass Spectrometry*, http://www.cfsan.fda.gov/~dms/clo4meth.html.

[43] Richter, B.E., Henderson, S., and Later, D., Accelerated solvent extraction (ASE) as a sample extraction technique for perchlorate in solid matrices, presented at PITTCON 2004, Chicago, IL, March 7-12, 2004.

[44] Niemann, R.A. and Krynitsky, A.J., Ion chromatographic analysis of food for perchlorate by suppressed conductivity compared to tandem mass spectrometry, presented at the 230th ACS National Meeting, Washington, D.C., August 28-September 1, 2005.

[45] Valentin-Blasini, L. et al., Analysis of perchlorate in human urine using ion chromatography and electrospray tandem mass spectrometry, *Analytical Chemistry*, 77, 2475, 2005.

[46] Thorne, P.G., Field Screening Method for Perchlorate in Water and Soil, Report ERDC/CRREL Technical Report 04-8, Applied Research Associates, South Royalton, VT, 2004.

[47] Strauss, S.H. et al., ATR-FTIR detection of less than or equal to 25 ug/L aqueous cyanide, perchlorate, and perfluorooctylsulfonate, *Journal AWWA*, 94, 109, 2002.

[48] Hebert, G.N. et al., Attenuated total reflectance FTIR detection and quantification of low concentrations of aqueous polyatomic anions, *Analytical Chemistry*, 76, 781, 2004.

[49] Neuhold, C.G. et al., Catalytic determination of perchlorate using a modified carbon paste electrode, *Analytical Letters,* 29, 1685, 1996.

[50] Ardakani, M. et al., Perchlorate-selective membrane electrode based on a new complex of uranil, *Analytical & Bioanalytical Chemistry,* 381, 1186, 2005.

[51] Ader, M. et al., Methods for the stable isotope analysis of chlorine in chlorate and perchlorate compounds, *Analytical Chemistry,* 73, 4946, 2001.

[52] Oak Ridge National Laboratory, Perchlorate treatment promising for double-duty, *ORNL Reporter,* Number 63, November 2004.

[53] Bao, H. and Gu, B., Natural perchlorate has a unique oxygen isotope signature, *Environmental Science & Technology,* 38, 5073, 2004.

[54] Gu, B. et al., Environmental forensics of perchlorate origin and biodegradation using stable isotope ratio analysis, presented at Partners in Environmental Technology Technical Symposium and Workshop, Washington, D.C., November 30-December 2, 2004, Poster Program Abstracts, 75.

[55] Hatzinger, P.B. et al., Determination of the origin and fate of perchlorate using stable isotope analysis, presented at Partners in Environmental Technology Technical Symposium and Workshop, Washington, D.C., November 30-December 2, 2004, Poster Program Abstracts, 86.

[56] Bohlke, J.K. et al., Perchlorate isotope forensics, *Analytical Chemistry,* 77, 7838, 2005.

[57] NASA, Advanced perchlorate-tracking technology to be applied to area groundwater by NASA team, *NASA Jet Propulsion Laboratory Bilingual Newsletter,* 3, 2005.

[58] Sturchio, N.C. et al., Chlorine isotope fractionation during microbial reduction of perchlorate, *Environmental Science & Technology,* 37, 3859, 2003.

[59] Coleman, M. et al., Stable isotope composition of perchlorate to monitor its biodegradation, attenuation by other processes, and fingerprint its source, presented at Partners in Environmental Technology Technical Symposium and Workshop, Washington, D.C., November 30-December 2, 2004, 82.

[60] Martinelangeo, P. K., et al., Gas-phase ion association provides increased selectivity and sensitivity for measuring perchlorate by mass spectrometry, *Analytical Chemistry,* 77, 4829, 2005.

[61] Kishi, T., Nakamura, J., and Arai, H., Application of capillary electrophore-

sis for the determination of inorganic ions in trace explosives and explosive residues, *Electrophoresis*, 19, 3, 1998.

[62] U.S. EPA, *Test Methods for Evaluating Solid Waste, Physical/Chemical Methods Draft Update IVA*, May 1998.

[63] U.S. EPA, 40 CFR Part 141, *Unregulated Contaminant Monitoring Regulation (UCMR) for Public Water Systems Revisions*, Proposed Rule, *Federal Register*, 70, 161, 49093, August 22, 2005.

[64] Environmental Data Quality Work Group (EDQW), Sampling and Testing for Perchlorate at DoD Installations, Interim Guidance, (available at http://www.navylabs.navy.mil/Archive/PerchlorateInterim.pdf), January 21, 2004.

[65] AFCEE Guidance for Contract Deliverables, Appendix C, Quality Assurance Project Plan, Final Version 4.0.01, May 2005.

[66] U.S. EPA, *Test Methods for Evaluating Solid Waste, Physical/Chemical Methods Draft Update IVB*, http://www.epa.gov/epaoswer/hazwaste/test/up4b.htm, November 2000.

[67] Yang, S-Y., EPA Office of Solid Waste, Methods developments activities for perchlorate ion in solids, presented at American Chemical Society Symposium Perchlorates: Science and Policy, August 2005.

[68] Eaton, A.D., Verifying the Reliability of EPA Method 314 to Measure Perchlorate at Sub ppb Levels vs New EPA Method Options, presented at the 14th Annual EPA Region 6 QA Conference, October 21, 2004.

[69] Wessling, E., Chemical Quality Remains Critical, presented at the 2005 DoD Environmental Monitoring and Data Quality Workshop, April 2005.

[70] Zeiner, S., Lahr, E., and Vitale, R., Perchlorate: Utilization of ion chromatography and liquid chromatography on characterization project, presented at the 22nd Annual Environmental Monitoring Conference, July 2005.

CHAPTER 4

How Does Perchlorate Behave in the Environment?

Perchlorate is highly soluble and stable under typical environmental conditions. As a result, it migrates readily via soil pore water, groundwater, and surface water migration. Perchlorate does not migrate through the vapor phase due to its low vapor pressure. Once released to the environment, the physical, chemical and biological processes that affect the fate and transport of perchlorate include dissolution of source material (in the case of solids), advection, dispersion and diffusion, sorption, and biological transformation. In addition, when perchlorate enters the environment through the disposal of concentrated brine wastes, density can drive the waste flow. Each of these processes is further discussed below.

Definitions
- Advection - Movement of a solute with the bulk flow of water
- Conservative tracer - Solute that does not interact with solid surfaces (i.e., porous media) or undergo chemical, biological or physical transformation
- Diffusion - Migration of a solute from an area of higher concentration to lower concentration
- Dispersion - Mixing due to migration of water at different velocities
- Electron acceptor - Chemical entity that receives an electron in oxidation-reduction reactions
- Electron donor - Chemical entity that provides an electron in oxidation-reduction reactions
- Evapotranspiration - Sum of evaporation and transpiration via plants
- Hydraulic conductivity - A measurement of the ability of geologic material to transmit water
- Hydraulic gradient - A measurement of the difference in water elevation between two points along a flow path
- Microcosm - A small system that represents a larger scale system
- Retardation coefficient - Ratio of the average groundwater velocity to the average solute velocity
- Sorption - Attraction to solid particles

4.1 Source Behavior

If perchlorate is not managed properly, it can be released to the environment in solid form, as in ammonium, sodium and potassium perchlorate salts, and can also be released to the environment as a liquid, such as in a concentrated brine or perchloric acid solution. If perchlorate is released to the environment as a solid, it will dissolve when in contact with water. The rate at which solid perchlorate will dissolve depends on the amount of surface area exposed, the amount of water present, temperature, and how long it is in contact with water.

In dry climates where evapotranspiration exceeds precipitation, perchlorate may not migrate appreciably into the soil column and may redeposit at or near the soil surface. In these environments, water may not penetrate the soil column very far before it reverses course and moves upward due to evaporation and transpiration processes. When water evaporates or is transpired, the previously dissolved perchlorate salts are left behind in the shallow soil column.

In contrast, in a humid climate leaching from solid perchlorate at the soil surface can be significant. The kinetic energy imparted to solid perchlorate salts by impacting raindrops could promote dissolution, as has been documented for the dissolution of individual explosives compounds from solid explosive mixtures [1]. Standing water at the soil surface in contact with solid perchlorate salts will also enhance dissolution. If solid perchlorate is released as a mixture that contains a less soluble component, and the less soluble component physically surrounds or otherwise inhibits contact of water with individual grains of perchlorate within the mixture, this may slow the dissolution process. This is likely the case when solid rocket fuel mixtures with polymeric binder materials are released to the environment.

In the case of perchlorate released to the environment in dissolved form, the perchlorate solution infiltrates into the soil surface, flows overland, or both. Depending on the environmental setting, a liquid release of perchlorate has significant potential to affect surface water and groundwater resources.

At some locations, such as manufacturing facilities and rocket motor washout facilities, highly concentrated perchlorate solutions or brines have been released. Density driven flow may control the movement of the brine in the subsurface. Density differences between the brine and ambient groundwater could lead to downward vertical migration of the brine with little influence by the prevailing groundwater flow regime. Like the migration of dense non-aqueous phase liquids (DNAPL) below the water table, brine pools may form on top of low permeability confining layers [2]. However, to date no case studies have documented the occurrence of perchlorate brine pools. It is possible that the perchlorate mass within the confining layer may act as a long-term source of groundwater contamination. As at DNAPL sites, release from these low permeability layers will be mass-transfer limited, which would limit the effectiveness of traditional pump and treat remediation methods.

4.2 Advection and Dispersion

Once dissolved in water, the flow rate and direction of the water largely controls the migration of perchlorate. The advective transport of dissolved perchlorate occurs anywhere perchlorate is dissolved in moving water, including unsaturated soils, in groundwater, and in surface water bodies, such as lakes and rivers. The rate of perchlorate migration will be similar to, but not exactly the same as, the average rate of water migration due to other physical processes collectively called dispersion. For example, when water migrates through soil, all of the water does not travel at exactly the same speed due to variations in flow paths. These variations in flow rate result in a mixing process that causes a plume to spread as it migrates. As a result, the center of mass of a dilute perchlorate plume will move at the same average velocity as groundwater, but the leading edge of the plume will migrate faster than the average groundwater velocity, as shown conceptually in Figure 4.1. Dispersion also occurs in surface water flow, when mixing occurs as a result of flow in the center of a river that is faster than the flow along the banks of the river.

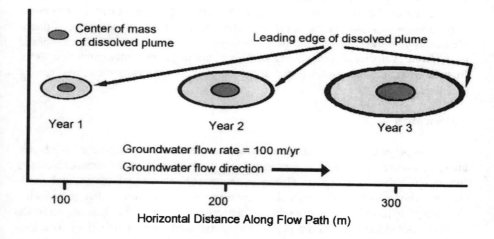

Figure 4.1   The Center of Mass of the Dissolved Plume Travels at the Same Average Velocity as Groundwater, While the Leading Edge of the Plume Migrates Slightly Faster

Estimating the rate at which perchlorate in groundwater will migrate is an important aspect of understanding potential impacts to resources such as a drinking water aquifer or an individual water supply well. The following groundwater velocity equation [3] can provide a simplistic estimate of the rate of perchlorate migration.

$$V = \frac{K}{n} \bullet \frac{dh}{dl}$$
(4-1)

Where:

V is the average linear velocity of groundwater,
K is the geologic material hydraulic conductivity,
n is the geologic material effective porosity, and
dh/dl is the hydraulic gradient

Then the travel time (t) to a particular resource at a distance (x) from the source can be estimated as:

$$t = \frac{x}{V} \qquad (4\text{-}2)$$

This simplified approach relies on the assumption that sorption and dispersion do not affect perchlorate transport. Based on the properties of perchlorate and characterization of perchlorate-contaminated sites reported to date, it is reasonable to ignore sorption. Given the level of confidence in the other parameters (some of which are often estimated to within no more than an order of magnitude), it is reasonable to ignore dispersion.

Considering the typical uncertainty in input parameters such as hydraulic conductivity, a range of input parameters should often be used to develop a range of possible perchlorate travel times. Further, this approach does not account for changes in groundwater velocity that may occur in the vicinity of groundwater extraction wells. In such cases, a more sophisticated approach such as numerical modeling may be beneficial.

4.3 Sorption

The sorption of ions on mineral surfaces depends on the degree of electronic charge on those surfaces. The presence of clay minerals or oxide surfaces, such as iron oxide, enhances sorption. However, the perchlorate anion does not sorb readily onto clay minerals: although clays have appreciable cation exchange capacity, they typically have little anion exchange capacity. With respect to binding on oxide surfaces, Schindler and Stumm [4] point out that weak bases (such as the perchlorate anion) typically form weak complexes on hydroxide surface sites. Thus, based on its charge and the even charge distribution that results from the tetrahedral symmetry, the perchlorate anion generally has a relatively low affinity for soils and mineral surfaces. It essentially behaves as a conservative tracer.

Studies performed at the University of California showed that adsorption of perchlorate did not appear to be significant in two different soils collected from agricultural fields in Yolo County, California. These soils consisted of 33 to 51% sand, 38 to 49% silt and 11 to 18% clay and also contained between 0.12 and 1.12% organic carbon. The study included both batch experiments and column experiments. The sterile batch experiments showed no indication of sorption of perchlorate. Saturated soil column experiments showed that the transport of perchlorate was very similar to bromide, which was used as an unreactive tracer. The column effluent concentrations were evaluated using an analytical solution of the advection-dispersion equation, which is a mathematical expression that predicts concentration as a function of time based on the bulk flow of water (advection), mixing due to flow in porous media (dispersion) and factors that slow the transport of a solute (retardation). The advection-dispersion equation was able to match the experimental data using a retardation coefficient equal to 1.0, which indicates that the migration of perchlorate in the experimental columns was not slowed by sorption to solid soil particles [5].

Sorption studies performed for the Kerr-McGee site in Henderson, Nevada are described below in a case study. The subsequent case study for the MMR site shows how the lack of sorption and biological attenuation can contribute to the development of an elongated dissolved plume of perchlorate, and describes the methods used to characterize that migration to support remedial actions.

Case Study: Kerr-McGee/PEPCON, Henderson, Nevada

The Kerr-McGee/PEPCON manufacturing facilities released perchlorate-containing wastes in liquid form to unlined settling ponds from the 1940s to 1976. The discharged wastes migrated through the unsaturated zone and formed perchlorate plumes in groundwater. The University of Nevada Las Vegas (UNLV) estimates that the plume emanating from the Kerr-McGee plant contains 20.4 million pounds of perchlorate dissolved in 9 billion gallons of water, and the plume emanating from the PEPCON plant has 1.1 million pounds of perchlorate in 9 billion gallons of water [6]. Contaminated groundwater has migrated toward and discharged to the Las Vegas Wash, which in turn discharges to Lake Mead and the Colorado River. The UNLV research team, which included members from the Desert Research Institute and Penn State, prepared a study for the U.S. EPA [6] that provides important information regarding the fate and transport of perchlorate. Their research included studies of sorption and biological degradation.

To evaluate the migration potential of perchlorate, the team collected soil and surface water samples along the Las Vegas Wash and also from Lake Mead. These samples were used in laboratory-based saturated column studies to evaluate the influence of soil sorption on the migration of perchlorate. Water containing dissolved perchlorate and bromide was fed through columns containing soil from the Las Vegas Wash. Bromide was used as a conservative tracer, to show the result of migration of a non-reactive solute through the soil columns, and estimate the column characteristics such as linear velocity and dispersivity. The effluent concentrations of perchlorate and bromide were very similar, which indicates that perchlorate behaved much the same as bromide in these column studies. The transport of perchlorate was modeled using a one dimensional transport model called CXTFIT. This analysis yielded perchlorate retardation coefficients ranging from 0.8 to 1.3. The authors concluded that these results were essentially equal to one, and that despite the relatively high specific surface area of soil from the Las Vegas wash, perchlorate behaves as a conservative tracer in these soils.

The research team also evaluated the biological degradation of perchlorate. The study isolated over thirty strains of perchlorate-reducing bacteria from soil and water samples collected from the Las Vegas Wash and Lake Mead. Despite the ubiquity and quantity of bacteria present in soil and water, microcosm studies found that perchlorate biodegradation was electron donor limited. In fact, in the absence of an external source of carbon, perchlorate biodegradation did not occur. In addition to the limitations posed by the lack of a carbon source, the study also showed that the presence of oxygen, nitrate and high salinity in the Las Vegas Wash limited perchlorate biodegradation. However, when an external source of carbon was added to samples where the salinity was not too high, biological reduction of perchlorate proceeded at a relatively rapid rate. These results suggest the potential for *in situ* biodegradation through the addition of organic carbon is high in areas where salinity is not too high.

Case Study: MMR, Cape Cod, Massachusetts

Demo 1 is a kettle hole that was used for demolition training and explosive ordnance disposal. Before soil remediation in 2004, the lithology beneath the Demo 1 kettle hole consisted of sand and clay in the top 7 to 10 ft, which then changed to mostly sand below 10 ft. The increased clay content near the ground surface contributed to ponding of surface runoff, which may have allowed for increased contact time with contaminants on the soil surface and promoted the dissolution of solid perchlorate particles.

Once in solution, perchlorate is relatively mobile and will leach through the vadose zone to groundwater, leaving relatively little residual contamination below the zone of particulate deposition and reworked soil. The flux of perchlorate from the unsaturated zone to groundwater resulted in a cigar-shaped plume approximately 9,200 ft long and 1,000 ft wide in the middle. Before groundwater remediation began in 2004, the plume contained approximately 100 pounds of perchlorate within 1.5 billion gallons of groundwater [7]. Perchlorate was detected in 15 of 72 (21%) soil samples with a maximum detected concentration of 29 $\mu g/kg$. The highest concentration of perchlorate in groundwater was 500 $\mu g/L$ at well MW-76M2, which is located 1,400 feet downgradient of the source area. Lower concentrations in source area groundwater (see figure to the right), and relatively low source area soil concentrations suggest the source of perchlorate is declining [7]. Perchlorate's high aqueous solubility and rapid transport away from the source have contributed to this decline.

The groundwater geochemistry at Demo 1 does not support biological degradation of perchlorate because the groundwater contains insufficient dissolved organic carbon to supply an electron donor, and the concentration of dissolved oxygen is too high to allow a reducing environment [8]. Based on the unsuitable conditions for biological transformation and perchlorate's minimal affinity for sorption, the primary factors influencing the migration of perchlorate in groundwater at Demo 1 are hydraulic gradients and subsurface lithology.

Aside from the minimal downward gradient directly below the Demo 1 depression, groundwater elevation data indicate there is largely horizontal flow all the way to Pew Road (some 6,400 ft downgradient), at which point the plume thins from over 100 ft thick to approximately 50 ft thick. In this area the plume lies above a clay and silt zone, which appears to be continuous in the direction perpendicular to flow. Although downward gradients are also evident in this area, the lack of perchlorate below these units suggests the lower permeability clays and silts slow contaminant migration to greater depths.

The figure to the right provides a vertical cross section along the perchlorate plume axis, and shows the relationship between the lithology and the distribution of perchlorate in groundwater [7].

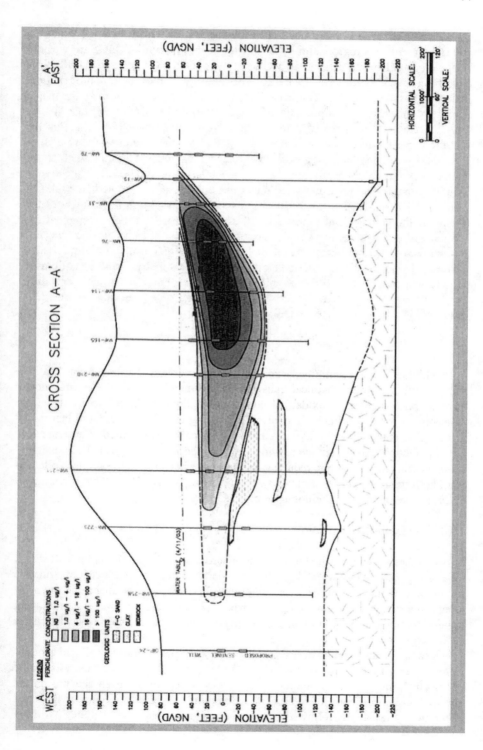

A numerical groundwater model was developed for Demo 1. This model, based on refinements to a regional model originally developed by the USGS for western Cape Cod, utilized tools such as MODFLOW, MODPATH and MT3DMS. In a hydrologic calibration, the hydraulic conductivity values in the model were adjusted based on local hydraulic conductivity information to optimize the match between the observed plume trajectory and travel times. The model also underwent a fate and transport calibration, which was accomplished by adjusting the source footprint and influent concentrations. This calibration assumed perchlorate does not biodegrade or adsorb to aquifer solids, and the longitudinal, transverse, and vertical dispersivity values were 3.00, 0.06 and 0.005 ft, respectively. The dispersivity values were based on values published by Garabedian et al. [9]. The ratio of longitudinal to transverse dispersivity in an aquifer influences the shape of the contaminant plume. The higher the ratio, the more elongated the plume will be along the direction of groundwater flow. The ratio used in the calibrated Demo 1 groundwater model was 50, which is higher than other published values [3], and consistent with the elongated plume geometry. The final calibrated model was able to match the observed perchlorate plume geometry reasonably well, and was successfully used to guide the design of a full-scale groundwater pump and treat system (see Chapter 7).

4.4 Biological Transformation

Perchlorate's low reactivity at ambient temperatures and dilute concentrations might imply that direct chemical transformation of perchlorate in the environment - through participation in oxidation-reduction reactions without the influence of microorganisms - is unlikely. However, a growing number of field and laboratory studies have shown that perchlorate can be biologically transformed. Bacteria capable of perchlorate degradation are widely distributed in nature [10] [11], and significant numbers of perchlorate reducing bacteria have been found in samples collected from both pristine and contaminated environments [6] [10] [12]. This section also discusses the uptake of perchlorate into plants.

4.4.1   Microbial Degradation

An increasing number of studies point toward a specific set of conditions that are required for natural biological degradation of perchlorate. There is a lack of evidence for biological transformation in aerobic environments, and in anaerobic environments, the presence of competing electron acceptors (e.g., chlorate, nitrate), and organic matter can be important controlling factors in perchlorate degradation, as described below.

Researchers have demonstrated the degradation of perchlorate in soil and groundwater under anaerobic conditions [13] [14] [15] [16] [17]. From a thermodynamic perspective, the order in which terminal electron acceptors are used depends on the energy yielding capacity of the electron acceptors. Typically oxygen is the most energetically favorable, followed by nitrate, manganese, iron and sulfate. Work by

Coates et al. [11] indicated that perchlorate reduction may yield about the same amount of energy as nitrate respiration, and suggested that excessive oxygen and nitrate may inhibit microbial attenuation of perchlorate.

More recent research at the University of California Davis found that nitrate reduction occurred much sooner than perchlorate reduction in soil with no history of exposure to perchlorate. In soils previously exposed to perchlorate, however, nitrate and perchlorate were simultaneously reduced at redox potentials ranging from 180 to 308 millivolts [16]. These measurements were made using three platinum redox electrodes, a combination electrode and a calomel reference electrode.

In contrast, work by Tan et al. [17] found that the presence of nitrate delayed perchlorate degradation in laboratory wetland column experiments, especially when the availability of organic material was a limiting factor. Tan et al. evaluated the influence of nitrate on perchlorate degradation in anaerobic sediments. Sediments from one site exhibited simultaneous perchlorate and nitrate reduction, but the results from four other sites indicated that the presence of nitrate prevents the use of perchlorate as an electron acceptor, and thus delays or prevents perchlorate degradation. The study also concluded that the presence of sulfate at 300 mg/L in the microcosm did not affect the degradation of perchlorate.

Tan et al. [18] then looked at the temporal and spatial variation of perchlorate in streambed sediments over an entire year at Naval Weapons Industrial Reserve Plant (NWIRP) in McGregor, Texas, by examining two stream segments continuously exposed to perchlorate and one intermittently exposed stream segment at six separate locations. In all sample locations, sediments at depths much below 20 cm did not contain detectable perchlorate. Perchlorate concentrations in the intermittently exposed stream segment showed temporal variability, with the greatest perchlorate penetration into sediments in warmer months and the least perchlorate penetration into sediments in the colder months. In one of the two continuously exposed stream segments, investigators did not observe perchlorate penetration into the sediments. The researchers noted that this segment had high plant growth, which may have provided higher organic substrate for perchlorate degradation. In addition, plant uptake of perchlorate may have reduced the amount of perchlorate reaching the sediments. The other continuously exposed stream segment had the highest concentrations of perchlorate in surface water. In this segment, perchlorate penetrated 10 to 20 cm into the sediments. Tan et al. attributable this finding, in part, to relatively high nitrate concentrations in surface water (10 mg/L) at this location, which may have inhibited perchlorate degradation. Overall, Tan et al. [18] concluded that the sediments were able to degrade perchlorate concentrations within the top 30 cm.

Several researchers have identified the availability of organic material as a crucial factor affecting biological degradation [6] [16] [17] [18]. Tan et al. [17] performed a series of laboratory experiments using sediment samples from streams impacted by perchlorate from the NWIRP, and the Longhorn Army Ammunition Plant in Karnac, Texas. Laboratory microcosm experiments started with an initial concentration of 5 mg/L perchlorate in the overlying water. Perchlorate degraded in all four studies; however, the authors found that the lag time for perchlorate degradation generally was longer under conditions of limited organic material.

First order degradation rates seemed to correlate with the amount of organic carbon, with higher perchlorate degradation rates in sediments with higher organic carbon content [17] [18]. The authors also noted that the prior exposure of sediments to perchlorate was less important than substrate availability. Table 4.1 summarizes the sediment type, organic carbon content and degradation rate data for each of three sediment types.

Table 4.1 Sediment Type and Perchlorate Degradation Rates for
McGregor Texas Site [17] [18]

Site Name	Sediment Type	Total Organic Carbon in Sediment (%)	Intrinsic First Order Degradation Rate (1/day)
HW317	Silty sand	16.05 ± 0.05	0.46
HW84 Side stream	Silty clay	11.59 ± 1.2	0.36
HW317/MN	Sandy clay	7.06 ± 0.02	0.13

The case study for the Indian Head Naval Surface Warfare Center illustrates the biological degradation of perchlorate.

Case Study: Indian Head Naval Surface Warfare Center, Indian Head, MD
Research performed at the Indian Head Division, Naval Surface Warfare Center (IHDIV, NSWC) demonstrates that environmental conditions may limit biodegradation of perchlorate despite the presence of perchlorate degrading bacteria. Researchers identified perchlorate-reducing bacteria in soil and water samples collected from three different locations at the IHDIV, NSWC. At one of these locations, Building 1419, a shallow, narrow plume of perchlorate-contaminated groundwater was identified where workers clean out or "hog out" solid propellant containing ammonium perchlorate from outdated materials such as rockets and ejection seat motors. Researchers identified low pH and a lack of sufficient electron donor as key factors limiting perchlorate biodegradation [19]. A subsequent field study demonstrated that naturally occurring organisms could be stimulated to degrade perchlorate *in situ* using lactate as a food source (electron donor) and a carbonate/bicarbonate mixture to buffer the pH [20]. Chapter 7 presents additional details on *in situ* biodegradation of perchlorate at Building 1419.

Another study performed at IHDIV, NSWC highlights two transport phenomena: first, storm water runoff can be a significant transport route for perchlorate, and second, perchlorate concentrations can be higher in plants than in co-located soil or sediment. Handling of perchlorate mixtures at the Low Vulnerability Ordnance Ammunition (LVOA) area resulted in migration of perchlorate in surface water through a series of creeks and marshes. Researchers detected perchlorate approximately 2,750 ft downstream of the LVOA in surface water and aquatic vegetation, as well as in soil and terrestrial vegetation. Terrestrial vegetation contained higher concentrations of perchlorate compared to other media, including soil collected at the same locations. Aquatic vegetation contained higher concentrations of perchlorate compared to other media, including sediment and surface water collected at the same locations [21]. Chapter 5 presents additional details on the exposure of aquatic and terrestrial organisms to perchlorate.

Releases to the environment through septic systems can also reflect microbiological transformations. In 2004, the Massachusetts Department of Environmental Protection performed a very limited study of the biodegradation of perchlorate entering two domestic septic systems from a tap water source [22]. Chemists used LC/MS/MS to analyze samples of the tap water and the effluent from the septic systems. Table 4.2 summarizes the data from this study.

Table 4.2 Degradation of Perchlorate in Two Septic Tanks [22]

| Location | Residence Time (hr) | Perchlorate Concentrations (μg/L) | |
		Tap Water	Effluent from Septic Tank
Condominium complex	< 48	783/943	0.23
Private residence	Unknown	190	< 0.2

These results are expected because septic tanks are anaerobic (i.e., oxygen is not present) and therefore perchlorate is a preferred electron acceptor. Furthermore, the organic material within septic tanks can be used by perchlorate reducing bacteria as an electron donor to fuel the reduction of perchlorate.

4.4.2   Uptake into Plants

Certain vascular plant species have been shown to take up perchlorate. Bulrushes, crabgrass, goldenrod, and cupgrass grown in perchlorate-contaminated areas have accumulated perchlorate. Crabgrass seeds contained perchlorate at a concentration of 1,880 mg/kg. For goldenrod, perchlorate concentrations were highest in leaves (1,030 mg/kg) but it also was present in stems, roots, and seeds [23]. Tobacco grown in soils treated with Chilean caliche fertilizers accumulated perchlorate in its leaves [24].

In some instances, higher plants can reduce perchlorate without the involvement of facultative anaerobic microorganisms [25]. In one experiment, researchers exposed hydroponically grown trees and plant nodules to ammonium perchlorate. The experiment occurred under sterile conditions to ensure that microbial activity did not contribute to perchlorate reduction. No toxic effects were noted and uptake was still occurring at the end of 30 days. Approximately 50% of the radioactive-labeled perchlorate was removed from solution by poplar trees and plant nodules in 30 days with 27.4% translocated in the leaves of the trees. Of the radioactivity remaining in solution, 68% remained as non-transformed perchlorate ion. Both in solution and in the leaves, labeled chlorine was associated with unmetabolized perchlorate, chlorate (ClO_3^-), chlorite (ClO_2^-) and chloride.

Willow trees have also proven capable of perchlorate degradation, with two distinctly different phyto-processes at work. Perchlorate can potentially be taken up and degraded in the leaves and branches; and degradation can occur in the rhizosphere by perchlorate-degrading microorganisms [26]. The presence of competing terminal electron acceptors, such as nitrates and other nutrients, interfered with the rhizodegradation of perchlorate.

Aquatic plants may also take up and accumulate perchlorate in various tissues. Bulrushes growing in ponds with perchlorate contamination at 30 to 31 mg/kg were found to accumulate perchlorate in tissues, 7 mg/kg, both above and below the waterline and in roots, with the highest accumulation in tissues above the waterline [23].

Section 5.2 provides additional discussion of perchlorate in plants, in particular from the perspective of exposure of terrestrial wildlife to perchlorate.

4.5 Summary

Perchlorate is highly soluble and stable under typical environmental conditions, and as a result it can move readily through the environment. Once released to the environment, the physical, chemical and biological processes that affect the fate and transport of perchlorate include dissolution of source material (in the case of solids), advection, dispersion and diffusion, sorption, and biological degradation. When dissolved in soil pore water, groundwater, and surface water, perchlorate's limited ability to sorb to mineral surfaces results in migration dominated by the bulk movement of water (advection) and mixing processes (dispersion). In cases where the advective velocity of the migrating water is slow, such as within a clay layer in an aquifer, diffusion processes may become important. Biological degradation of perchlorate has been demonstrated in field and laboratory studies under anaerobic conditions, and generally requires the presence of an electron donor, such as naturally occurring organic material, and sufficiently low concentrations of oxygen and nitrate. The presence of oxygen or nitrate can inhibit or delay the reduction of perchlorate. Terrestrial and aquatic plants may also take up perchlorate and accumulate it in various tissues. Perchlorate can be taken up and degraded in the leaves and branches while degradation can occur in the rhizosphere by perchlorate-degrading microorganisms.

4.6 References

[1] Lynch, J.C., Dissolution kinetics of high explosive compounds (TNT, RDX, HMX), ERDC\EL TR-02-23, *U.S. Army Corps of Engineers, Engineer Research and Development Center, Environmental Laboratory,* Vicksburg, MS, 2002.

[2] Flowers, T.C. and Hunt, J.R., Long-term release of perchlorate as a potential source of groundwater contamination, in *Perchlorate in the Environment,* Urbansky, E.T. Ed., Kluwer/Plenum, New York, 2000, Chap. 17.

[3] Fetter, C.W., Contaminant Hydrogeology, Prentice Hall, Upper Saddle River, NJ, 1993, p. 47, 66.

[4] Schindler, P.W., and Stumm, W., The Surface Chemistry of Oxides, Hydroxides and Oxide Minerals in *Aquatic Surface Chemistry*, Stumm, W. Ed. Wiley Interscience, New York, 1987, Chap. 4.

[5] Tipton, D.K, Rolston, D.E. and Scow, K.M., Bioremediation and biodegradation, transport and biodegradation of perchlorate in soils, *J. Environ. Qual.* 32:40-46, 2003.

[6] Batista, J.R. et al., Final Report: The Fate and Transport of Perchlorate in a Contaminated Site in the Las Vegas Valley, Prepared for the United States Environmental Protection Agency, 2003.

[7] AMEC, Impact Area Groundwater Study Program, Final Feasibility Study, Demo 1 Groundwater Operable Unit, Prepared for U.S. Army Corps of Engineers and U.S. Army/National Guard Bureau, August 19, 2005, p. 3-7.

[8] AMEC, Final IAGWSP Technical team memorandum 01-2, Demo 1 ground-water report for the Camp Edwards Impact Area Groundwater Quality Study, Massachusetts Military Reservation, Cape Cod Massachusetts, prepared for the National Guard Bureau, Arlington, VA and the Massachusetts Army National Guard by AMEC Earth and Environmental, 2001.

[9] Garabedian, S.P et al., Large scale natural gradient tracer tests in sand and gravel, Cape Cod, Massachusetts - 2. Analysis of spatial moments for a nonreactive tracer, *Water Resources Research,* 27, No. 5, p. 911-924, 1991.

[10] Coates, J.D. et al., Ubiquity and diversity of dissimilatory (per)chlorate-reducing bacteria, *Appl. Environ. Microbiol.,* 65, p. 5234-5241, 1999.

[11] Coates, J.D., The diverse microbiology of (per)chlorate reduction, in *Perchlorate in the Environment,* Urbansky, E.T. Ed., Kluwer/Plenum, New York, 2000, Chap. 24, p. 257-270.

[12] Wu, J., et al., Persistence of perchlorate and the relative number of perchlorate- and chlorate-respiring microorganisms in natural waters, soils, and wastewater. *Biorem. J.,* 5, 119-130, 2001.

[13] Attaway, H. and Smith, M., Reduction of perchlorate by an anaerobic enrichment culture, *J. Ind. Microbio.,* 12, 408-412, 1993.

[14] Rikken, G.B., Kroon, A.G., and Van Ginkel, C.G., Transformation of (per)chlorate into chloride by a newly isolated bacterium: Reduction and dismutation, *Appl. Microbial. Biotechnol.,* 45, 420-426, 1996.

[15] Logan et al., Kinetics of perchlorate- and chlorate-respiring bacteria, *Appl. Environ. Microbiol.,* 67, 2499-2506, 2001.

[16] Tipton, D.K., Rolston, D.E. and Scow, K.W., Bioremediation and biodegradation: transport of biodegradation of perchlorate in soils, *J.Environ. Qual.*, 32, 40-46, 2003.

[17] Tan, K., T.A. Anderson, and W.A. Jackson, Degradation kinetics of perchlorate in sediments and soils, *Water, Air, and Soil Pollution*, 151: 245-259, 2004.

[18] Tan, K., Anderson, T.A., Jackson, W.A., Temporal and spatial variation of perchlorate in streambed sediments: results from in-situ dialysis samplers, *Environ. Pol.*, 136, 283-291, 2005.

[19] Hatzinger, P.B., In situ bioremediation of perchlorate, Final Report for SERDP Project CU-1163, Strategic Environmental Research and Development Program, Arlington, VA, p. 131, 2002.

[20] Cramer, R.J., et al., Field demonstration of in situ perchlorate bioremediation at building 1419, Prepared for Naval Ordnance Safety and Security Activity, Ordnance Environmental Support Office, NAVSEA Indian Head, Surface Warfare Center Division, NOSSA-TR-2004-001, January 2004.

[21] Parsons Engineering Science, Inc., Scientific and technical report for perchlorate biotransport investigation: a study of perchlorate occurrence in selected ecosystems, Interim final, Austin, TX, contract no. F41624-95-D-9018, 2001.

[22] Massachusetts Department of Environmental Protection, *The Occurrence and Sources of Perchlorate in Massachusetts DRAFT REPORT* August 2005, http://www.mass.gov/dep/cleanup/sites/percsour.pdf, p. 31-39.

[23] Smith, P.N., et al., Preliminary assessment of perchlorate in ecological receptors at the Longhorn Army Ammunition Plant (LHAAP) Karnack, Texas, *Ecotoxicology*, 10, 305-313, 2001.

[24] Ellington, J.J., et al., Accumulation of perchlorate in tobacco plants and tobacco products, *Environmental Science and Technology*, 35, 3213-3218, 2001.

[25] Van Aken, B. and Schnoor, J.L., Evidence of perchlorate (ClO_4^-) reduction in plant tissue (poplar tree) using radio-labeled $^{36}ClO_4^-$, *Environmental Science and Technology*, 36, 2783-2788, 2002.

[26] Nzengung, V., Wang, C., and Harvey, G., Plant-mediated transformation of perchlorate into chloride, *Environmental Science and Technology*, 33, 1470-1478, 1999.

CHAPTER 5

What Are the Implications of Human and Ecological Exposures to Perchlorate?

"All substances are poisons; there is none which is not a poison. The right dose differentiates a poison and a remedy." (Paracelsus c.1493-1541)

With as much certainty as the science currently offers, this chapter focuses on the question of whether, and to what degree, the presence of perchlorate in the environment poses a threat to human health and the environment. Perchlorate compounds find many uses and can originate from natural sources. This chapter describes how humans and other species may be exposed to possible environmental sources such as soil, surface water, and groundwater, and the current knowledge of the toxic properties of perchlorate.

Definitions
- Acute toxicity - Adverse effects manifested within a relatively short time interval (minutes to days).
- Ambient Water Quality Criteria - Levels of pollutants in surface water, estimated using U.S. EPA methods, considered to be protective of aquatic organisms.
- Bioaccumulate - The net accumulation, over time, of a substance within an organism obtained from both biotic (prey species) and abiotic (soil, sediment, and water) sources.
- Bioconcentrate - An increase in concentration of a substance in an organism to levels greater than in the surrounding medium.
- Bioconcentration factor - The ratio of the concentration of a compound in an organism to the concentration in the surrounding medium (e.g., perchlorate in plant vs. perchlorate in soil).
- Carcinogenic - An agent or event that modifies the genome and/or other molecular control mechanisms of the target cells, giving rise to a population of altered cells.
- Chronic toxicity - The adverse effects expressed after long period of exposure to small quantities of a toxicant.
- Clinical Study - A study in which patients or volunteers are studied to determine the effect, safety or efficacy of a compound, such as a therapeutic agent.
- Epidemiological Study - The study and investigation of the distribution and cause of disease(s).
- Expected Environmental Concentration (EEC) - The predicted concentration

of a pollutant in one or more media (e.g., soil, surface water, sediment, air), typically calculated from a real or anticipated release scenario.

- Genotoxic - An adverse effect resulting in an alteration of the genome of a living cell. The alteration can be expressed as a mutagenic or carcinogenic event.
- Half-life - The time period necessary for one-half of the substance to disappear from an organism or environmental medium.
- Hazard - The qualitative nature of the adverse effect resulting from a particular toxic chemical, physical effect, or inappropriate action.
- Hazard quotient - The ratio of estimated site-specific exposure to a single chemical to the estimated safe daily exposure level (at which no adverse health effects are likely to occur).
- Lowest Observed Effect Concentration - In a toxicity test, the lowest concentration at which a statistically or biologically significant effect is observed in the exposed population compared with an appropriate unexposed control group.
- Mutagenic - Capable of producing a mutation, which is an alteration in the genetic material (DNA) of living cells.
- No Observed Effect Concentration - In a study toxicity test, the highest concentration of a chemical that causes no observable effect.
- Receptors - Group of individuals who may be exposed to environmental contamination.
- Reference dose - An estimate (with uncertainty spanning perhaps an order of magnitude) of a daily oral exposure to the human population, including sensitive subgroups, that is likely to be without an appreciable risk of deleterious effects during a lifetime.
- Risk - The probability that some adverse effect will result from exposure to a chemical.
- Risk assessment - The science-based process by which potential adverse health effects of exposure to chemicals are characterized.
- Risk management - The process of determining how to regulate or mitigate a risk by considering social, economic, and political implications of such options.
- Toxicity Reference Value (TRV) - Benchmark value representing dose to the specific receptor that is unlikely to cause adverse effects with chronic exposure.
- Uncertainty factor - A numeric value intended to account for the uncertainties in deriving an acceptable dose of a chemical. Also referred to as a Safety Factor.

5.1 Overview of Potential Human Health Effects and Components of Risk Assessment

State or federal environmental authorities determine if a substance is sufficiently hazardous to require management or regulation in a formal process called risk assessment. That process estimates the probability that some harmful event could occur under prescribed conditions. For example, a risk assessment at a Superfund site might conclude that a lifetime of drinking contaminated water containing a carcinogen might result in one case of cancer in 10,000 people (1×10^{-4} excess cancer risk). Regulators must then make a risk management decision as to whether that level of risk based solely on the risk assessment is acceptable or unacceptable in the context of social, political, personal and economic factors. Thus, the regulations ultimately derived to protect human health and the environment from perchlorate or other environmental contaminants requires both scientifically-based risk assessments and other considerations as part of the risk management decision process. The focus of this chapter is on the risk assessment process while Chapter 6 provides an understanding of how the conclusions from the risk assessment are considered in the context of all the other factors to regulate perchlorate in the risk management process.

The risk assessment process has become part of our modern life in a multitude of applications. We see the results of risk assessment expressed in association with various familiar activities, such as the risk of dying from lung cancer for cigarette smokers versus non-smokers (three times as likely) [1], or the chance of severe head injury when riding a bicycle without a helmet (bicycle helmets can reduce the risk of a head injury by 85 percent and brain injury by 88 percent [2]). Such risks are calculated by actuaries and epidemiologists based on population statistics.

The goal of an environmental risk assessment is to estimate the extra risk caused by the exposure to an environmental contaminant over the risk that exists when the chemical exposure does not occur. Environmental risk assessments produce estimates of this incremental risk or additional hazard, not absolute projections of effects. Although the risks from exposure to chemicals in the environment, such as perchlorate, are expressed in a similar manner to activity risks, much of the information regarding such exposures is less certain. To account for these unknowns, scientists estimate the risks from exposure to environmental contaminants using extrapolations beyond the measured risk information (also referred to as "uncertainty factors"). Finally, although the term "risk" carries a connotation of causation, a risk assessment does not ascribe actual exposures with absolute adverse or harmful effects.

Scientists assessing the possible risks from chemical exposures consider different types of information about the chemical, including possible sources in environmental media (e.g., natural or anthropogenic), the toxicity of the chemical and whether the chemical is considered a human carcinogen, scientific studies of the nature of the toxicity, if any, and conditions of possible exposure. The National Research Council (NRC) [3] of the National Academy of Sciences (NAS) developed a four-component risk assessment paradigm that the U.S. EPA adopted [4] to characterize potential risk to human health from exposure to contaminants at hazardous waste sites. These four components are:

- Hazard Identification - In this step, risk assessors summarize the contaminants one might be exposed to in various environmental media. This includes: (1) compiling data on the concentration and distribution of a contaminant within the different environmental media, such as soil or water used as a source of drinking water, (2) itemizing all the contaminants found at a site within specific chemical classes (e.g., metals, pesticides, volatile organic compounds), and (3) summarizing the nature of the toxicity of the contaminants (that is, whether it is considered a possible human carcinogen or may have toxic non-carcinogenic effects). The Hazard Identification forms the basis for the next three steps in the risk assessment process.

- Dose-response assessment - This step of the risk assessment documents the current scientific understanding of the harmful effects (or toxicity) of the contaminants documented in the Hazard Identification step. The term "dose-response" describes the relationship between the amount of the contaminant administered or received (referred to as "dose") and the incidence of adverse health effects in the exposed population (or "response"), usually assessed by conducting animal studies. For each compound, the dose-response relationship serves as the basis for the toxicity values that are used to estimate the potential incidence of adverse health effects in an exposed population. The toxicity values for effects other than cancer are termed reference doses. The toxicity values for carcinogenic effects are termed cancer slope factors.

- Exposure assessment - This step uses the information from the Hazard Identification and determines how, and to what degree, people (also referred to as "receptors") are potentially exposed to contaminants at the site. In addition, risk assessors consider whether future exposures could be the same or different from current exposures based on the chemical properties of the contaminant, the medium where it is found, and the effects of weathering (referred to as "fate and transport changes").

- Risk characterization - The final step of the risk assessment combines the results of the exposure assessment with chemical-specific toxicity information to quantify potential risks. Risk managers then use these risk estimates to decide what, if anything, to do about the situation.

The remainder of this section, which discusses the potential threats to human health from exposures to perchlorate in the environment, follows this four-step paradigm for risk assessment.

5.1.1   Hazard Identification

The purpose of a Hazard Identification is to determine where contamination may exist and at what levels and frequency of detection, what types of receptors may be at risk (e.g., occupational, elderly, children, etc.), and the potential that a given population may be exposed to the contamination. The Hazard Identification is highly site-specific.

It has been nearly ten years since the U.S. EPA embarked on a fairly rigorous evaluation of the exposure and subsequent health effects of perchlorate in humans.

Improvements to analytical methods in 1997 allowed scientists to identify low levels of perchlorate in groundwater at hazardous waste sites, and in some public water supplies (often downgradient of hazardous waste sites or military operations). It is currently estimated that over 11 million people have perchlorate in their public drinking water supplies at concentrations of at least 4 μg/L [5]. It is also known, however, that the vast majority of detected concentrations in U.S. water supplies are less than 12 μg/L [6]. For the general population, including sensitive subpopulations as discussed further in Chapter 6, the U.S. EPA considers 24.5 μg/L an acceptable level in drinking water [7].

5.1.2   Dose-Response Assessment

Researchers have produced voluminous information on the potential health effects of perchlorate. All of this information has been exhaustively reviewed and published in several comprehensive documents. As of this writing, the U.S. EPA is finalizing a draft publication entitled *Perchlorate Environmental Contamination: Toxicological Review and Risk Characterization* [8] which addresses every facet of the known health implications following exposure to perchlorate. In addition, the Agency for Toxic Substances and Disease Registry (ATSDR) has published a *Draft Toxicological Profile on Perchlorates* [9]. Finally, at U.S. EPA's request, the National Research Council performed a detailed scientific review and expert opinion reported in *The Health Implications of Perchlorate Ingestion* [5]. Given the explosion of information on nearly every aspect of the perchlorate issue (analytical, occurrence, fate, transport, regulatory, health, waste treatment, etc.), it is not possible to cover all publications and related scientific journal articles in this chapter. This discussion therefore focuses on the highlights of the NRC report [5], which forms the current basis for the U.S. EPA reference dose [10].

The NRC "assess[ed] the current state of the science regarding potential adverse effects of disruption of thyroid function by perchlorate in humans and laboratory animals at various stages of life" and "evaluate[d] the animal studies used to assess human health effects of perchlorate ingestion with particular attention to key end points, including brain morphometry, behavior, thyroid hormone levels, and thyroid histopathology." The NRC found numerous human studies available and that the quality of these data sufficed to derive a protective, health-based value for the ingestion of perchlorate. The NRC determined that the use of animal testing data was therefore not only unnecessary but, once the animal bioassay data were reviewed, observed several discrepancies (e.g., poor histological methods, U-shaped dose response curve) that threw some of the data into question.

Primer on Thyroid Function [5] [11]

The thyroid gland is a pinkish-red butterfly-shaped organ attached to anterior cartilage of the trachea. It comprises millions of spherical follicles that contain a colloid-like protein, thyroglobulin. The cuboid-like cells surrounding these follicles manufacture, store and secrete the hormone thyroxine, a compound synthesized from tyrosine that contains iodine (T_3 and T_4 contain 3 and 4 molecules of

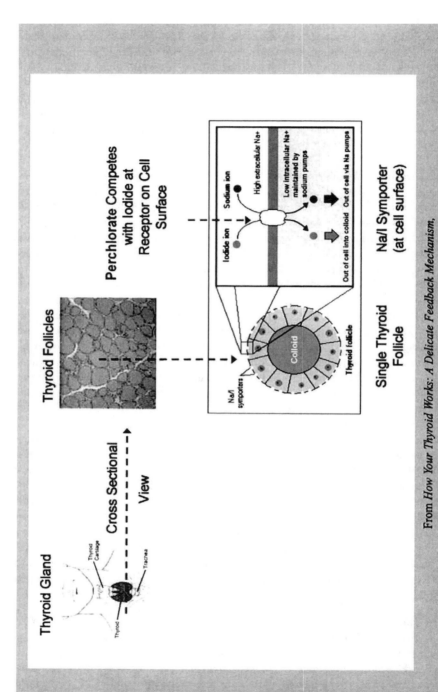

From *How Your Thyroid Works: A Delicate Feedback Mechanism*, Norman Endocrine Surgery Clinic, www.endocrineweb.com/thyfunction.html. [11] With permission. Histology graphic courtesy of Dr. Roger C. Wagner, University of Delaware: http://www.udel.edu/Biology/Wags/histopage/histopage.htm. [80]

iodine, respectively). The production of thyroxine, in turn, is regulated by the pituitary gland at the base of the brain. Thyroxine functions to regulate metabolism in warm-blooded animals. Because all cells need to metabolize nutrients, the thyroid gland is also very important during the development and maintenance of many different types of organs in the body.

The interaction between the pituitary gland, which secretes thyroid-stimulating hormone (TSH) and the thyroid gland, which secretes T_3 and T_4, is known as a feedback loop. It has been compared to a thermostat and a furnace, whereby an increase in the room temperature will trip the thermostat, which shuts down the activity of the furnace. When the level of thyroid hormones (T_3 & T_4) drops too low, the pituitary gland produces TSH which stimulates the thyroid gland to produce more hormones. Under the influence of TSH, the thyroid will manufacture and secrete T_3 and T_4 thereby raising their blood levels. The pituitary senses this and responds by decreasing its TSH production.

The thyroid follicle cells take up iodine from the bloodstream. This uptake process is regulated by another protein, known as the sodium-iodide (Na+/I-) symporter, that is embedded into the outer surface of the thyroid follicle cells. Because the size and charge of perchlorate is similar to that of iodine, it will block the Na+/I- receptor and keep iodide from entering into the lumen of the thyroid follicle. However, because this inhibition of the uptake by perchlorate is reversible, and because perchlorate is effectively cleared by the kidneys, it takes a significant dose over a fairly long period of time for perchlorate to cause a deficiency of iodine (a condition known as hypothyroidism).

Following their review of all of the relevant studies, the NRC decided to focus only on human studies and concluded the following with regard to the identification of hazards of perchlorate to the human population [5]:

- Perchlorate is a fully oxidized monovalent anion which is highly water soluble. As a result it does not react with human tissues (i.e., it cannot be metabolized to another state or compound). Once ingested, it is rapidly absorbed, has a short half-life in the human body (~8 hours) and is rapidly excreted unchanged in the urine.
- The only known effect of low level perchlorate exposure in humans is the competitive (reversible) inhibition of the uptake of iodide by the thyroid gland. It was therefore used in the 1950s and 1960s as an effective drug to treat hyperthyroidism. Treatment resulted in few side effects, the frequency of which depended on the dose.
- No evidence suggests that perchlorate causes thyroid disorders, thyroid nodules or cancer in the thyroid gland or any other organ. It is also not genotoxic or mutagenic.
- Perchlorate has been given to healthy volunteers, with doses ranging from 0.007-9.2 mg/kg•day, for up to six months. The lowest level that did not significantly inhibit the uptake of radioactive iodide was 0.007 mg/kg•day. There were no associated changes in serum thyroid hormones or thyroid-

stimulating hormone (TSH) to suggest thyroid hormone production was adversely affected.

- To cause a decline in thyroid hormone production that would have adverse health effects, iodide uptake would most likely have to be reduced by at least 75% for several months or longer.
- The dose of perchlorate required to cause hypothyroidism in adults would probably be more than 0.4 mg/kg per day, assuming a 70 kg body weight.
- Data from laboratory animal studies are limited in their usefulness for quantitatively assessing human health risk associated with perchlorate exposure.

To date, there is only one known effect of low level perchlorate exposure in humans, and that is the inhibition of the active uptake of iodide by the thyroid gland. The perchlorate ion blocks the sodium-iodide symporter (NIS) protein that normally acts as an iodide pump on the surface of the thyroid follicle. This is a competitive inhibition and therefore reversible when exposure to perchlorate ceases. Perchlorate is, therefore, not a toxic compound *per se* because it does not cause a direct toxic effect via a mechanism that can cause injury to a cell or tissue. The only adverse effect is, ultimately, hypothyroidism, which is an indirect effect caused by prolonged exposure to perchlorate. A visible swelling of the thyroid, or goiter, would occur before adverse effects from iodide deficiency would occur. In other words, a person would see a visible enlargement of the thyroid gland in the neck before experiencing symptoms of hypothyroidism (e.g., weakness, fatigue, weight gain, coarse dry hair and skin, chills, irritability, etc.).

Persons who work at perchlorate manufacturing facilities would be expected to have a high potential for exposure to perchlorate relative to the general population. Table 5.1 summarizes four recent occupational studies that examined clinical aspects of thyroid function in workers exposed to high concentrations of perchlorate in the air, with body burdens ranging from 0.3 to 38 mg/kg [5]. With the exception of the study which evaluated the rate of cancer mortality, all of these studies measured thyroid hormone endpoints that might be expected to change following inhalation of perchlorate. None of these studies showed any significant change in biomarkers associated with exposure to perchlorate. Although each of these studies has some experimental design deficiency, the weight of evidence strongly suggests the conclusion that no adverse effects occurred despite exposure to relatively high occupational concentrations of perchlorate.

Other studies of the effects of perchlorate exposure in humans included voluntary clinical experiments or involuntary ingestion of perchlorate in populations that consume drinking water containing perchlorate. Table 5.2 presents the key clinical and epidemiological human studies identified by the authors of the NRC report [5]. These studies measured sensitive biomarkers of thyroid dysfunction (e.g., TSH, T_4, T_3, radioactive iodide uptake) in infants, mothers and/or adults. As discussed in the NRC report, each study has one or more deficiencies that needs to be taken into consideration when interpreting the results. In general, one can observe that the majority of these studies examining the effect of perchlorate exposure to humans

Table 5.1 Summary of Occupational Exposure Studies Addressing the Human Effects of Perchlorate

Number of Participants or Volunteers	Exposure Duration	Estimated Exposure or Dose	Major Endpoint(s)	Bias or Modifying Factors	Significant or Adverse Effect?	Ref.
3,963	Work Shift	Unknown	Cancer Mortality Rate	Most workers male; multiple chemical exposures.	No (no significant increase in cancer rate due to perchlorate exposure)	[15]
100	Work Shift	0.2-436 µg/kg	Change in TSH, thyroid hormone status	Small number of participants with some bias due to turnover.	No	[16]
289	Lifetime	3.5-38 mg/kg	Change in TSH, thyroid hormone status	No statistical adjustment for Body Mass Index or activity level; some turnover of participants.	No	[16]
52	Work Shift (varies w/age, task)	1, 4, 11 and 34 mg/shift	All thyroid hormone endpoints; urine iodide and creatinine	Small numbers of participants with some bias due to turnover.	No	[17]
29	Work Shift	0.3 mg/kg	Serum TSH, T3, T4, free T4 Index, radioactive iodide uptake, urinary iodide, serum perchlorate, thiocyanate and nitrate, thyroid size	Small number of non-exposed volunteers (control); pre- and post-shift evaluation allows for comparison within the same individuals; data are "preliminary."	Yes (decrease in mean radioactive iodide uptake post-shift; TSH unchanged, T3, T4, free T4 Index had small post-shift increases)	[18]

Table 5.2 Summary of Clinical and Epidemiologic Studies [5]

Number and Age of Study Participants	Exposure Type or Duration	Estimated Exposure or Dose	Major Endpoint(s)	Estimated "Safe" Concentration (µg/L) or Dose (mg/kg·day)	Ref.
5 Adults	28 Days	13 mg/kg·day	Change in radioactive iodide uptake	N/A: LOAEL 13 mg/kg·d	[19]
700,000 Newborns	Gestation [1]	4-16 µg/L and 5-8.3 µg/L	Congenital Hypothyroidism	4 µg/L	[20]
23,190 Newborns	Gestation [1]	9-15 µg/L; 0.9-4.2 mg maternal dose	Mean T4	15 µg/L	[21]
540 Newborns	Gestation [1]	9-15 µg/L; 0.9-4.2 mg total maternal dose	Mean TSH	15 µg/L	[22]
162 6-8 yr olds; 9,784 newborns	Chronic; Gestation [1]	<5, 5-7, and 100-120 µg/L	Serum TSH; free T3 and T4; clinical goiter	<5 µg/L	[12]
7-10 adults per dose group	14 Days	0.007, 0.1, and 0.5 mg/kg·d	Change in radioactive iodide uptake	0.7 µg/L	[23]
9 Adults	14 Days	0.14 mg/kg·d	Change in radioactive iodide uptake	N/A; NOAEL 0.04 mg/kg·day	[24]
507,982 Newborns	Gestation [1]	1-2 µg/L, 3-13 µg/L and >12 µg/L	Blood T4, TSH & congenital hypothyroidism	Effects seen at low and medium exposures, but not high	[25]
Low T4 infants born between 10/94 and 12/97	Gestation [1]	ND and <6 µg/L	TSH	6 µg/L	[26]
<18 yrs with diagnosis of ADHD or autism on Medicaid	Chronic	23.8 µg/L (median)	ADHD or autism: national percentile 4th grade test scores	No Effect Reported	[27]

Table 5.2 Summary of Clinical and Epidemiologic Studies, continued [5]

Number and Age of Study Participants	Exposure Type or Duration	Estimated Exposure or Dose	Major Endpoint(s)	Estimated "Safe" Concentration (µg/L) or Dose (mg/kg-day)	Ref.
Newborns	Gestation[1]	ND and <6 µg/L	TSH	6 µg/L	[28]
711,860 Newborns	Gestation[1]	< 9 µg/L (Mean = 1 µg/L)	TSH & congenital hypothyroidism	No Effect Reported	[29]
164 Mothers & Newborns	Gestation[1]	0.46, 5.8 and 114 ug/L	TSH, T3 & T4	<114 µg/L	[13]
342,257 Newborns	Gestation[1]	>5 and <5 µg/L	TSH & congenital hypothyroidism	> 5 µg/L	[14]

Notes:
ADHD = Attention Deficit Hyperactivity Disorder
ND = Not Detected above laboratory reporting limit
DW = Drinking Water

N/A = not applicable (or cited)
TSH = thyroid-stimulating hormone
(1) Maternal Exposure via drinking water intake during gestation

are negative or show no significant adverse effects. In some cases, such as the exposure of mothers and their newborns to naturally occurring perchlorate in drinking water in Chile [12] [13] or mothers and their newborns in California exposed to man-made sources of perchlorate in drinking water [14], the exposures are not only chronic, but are well above acceptable health standards for drinking water such as the U.S. EPA's Drinking Water Equivalent Level of 24.5 μg/L [7]. The latter study [27] is particularly notable because of the large population of newborns (more than 340,000) that the study had evaluated, which is the hallmark of a robust statistical analysis.

5.1.3   Exposure Assessment

The conditions of human exposures to perchlorate depend in part on the uses and properties of perchlorate, both presented in Chapter 2. These, in turn, determine if and how perchlorate could cause injury to human beings. Figure 5.1 illustrates how exposure to perchlorate can occur.

Once released to the environment, perchlorate readily dissolves in rainwater where it can migrate to surface water or percolate through the soil to the underlying groundwater. The resulting concentration of perchlorate in surface water or ground-water depends upon the amount of perchlorate initially released or found in soil (including the frequency of events that lead to the release of perchlorate), the nature of the soil (e.g., non-porous rock or sand), the amount of rainfall over time, and the distance to surface water or groundwater. In areas where the depth to groundwater is large and recharge rate is low, perchlorate may only migrate to subsurface soils and never reach the underlying groundwater aquifer. Once in an aqueous medium such as groundwater or surface water, perchlorate is very stable due to its solubility and persistence in the ionic state.

In general, humans are exposed to thousands of man-made and natural chemicals, whether as part of the food, nutrients, and medicines we consume, cosmetics we apply to our bodies, or work place and home environments in which we live. Exposure occurs through ingestion and inhalation, absorption through our skin or eyes, as well as other ways such as injection of certain medicines/drugs.

Medicinal Use of Perchlorate

Perchlorate is one of only a handful of chemical compounds where the true "mechanism of action" is known, that mechanism being the competitive inhibition of the uptake of iodide by the thyroid gland. In the 1950s and early 1960s, physicians took advantage of this known iodide blocking effect and used potassium perchlorate as a drug to combat hyperthyroidism. This is also known as Grave's Disease, a condition where the thyroid becomes overactive, resulting in life-threatening symptoms such as a racing heart, sweating, nervousness, loss of weight and sleeplessness. Patients treated with perchlorate (typically ~35 to 50 mg/day) for several months, would see a relief of symptoms and a return to normal life. Some patients, however, had side effects that may have included skin rashes, nausea and vomiting. Other patients, treated with very high levels of perchlorate over long

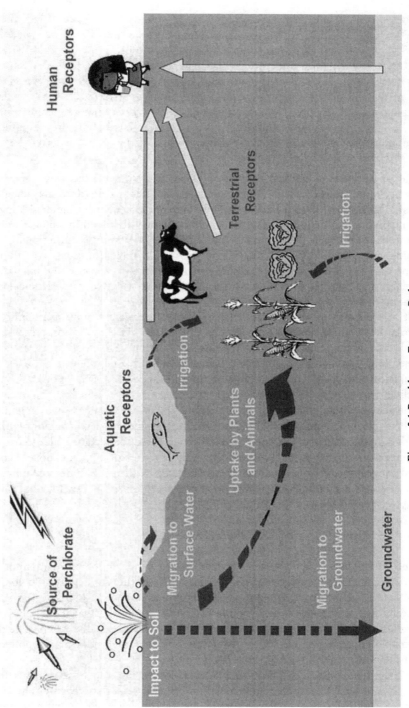

Figure 5.1 Perchlorate Exposure Pathways

periods of time (up to 400 mg/day), developed anemia (a shortage of red blood cells), and a few even died. Although concerns over the side effects prompted the development of new drugs and thus a decrease in the use of perchlorate (within the U.S.) as an anti-thyroid drug, it is still administered to Grave's Disease patients in some European countries. Perchlorate is still available as a drug in the United States, but principally to block the uptake of a radioactive compound containing technetium (e.g., $^{99}TcO_4^-$) into the thyroid following medical imaging tests of the brain, blood or placenta. It is also used to combat the thyroid effects of amiodarone, a drug used to treat heart arrhythmia [5] [9].

The physical and chemical properties of perchlorate limit the ways in which humans can be exposed to it. Due to its solubility, perchlorate does not generally persist in soil except in arid climates. This limits the potential for incidental ingestion of soil containing perchlorate. Furthermore, once dissolved in water, perchlorate exists as an ion, which limits its ability to penetrate the skin during bathing or washing. Because of its very low vapor pressure, perchlorate is not volatile and cannot exist as a vapor in air, resulting in negligible exposure by inhalation. However, inhalation of perchlorate in shower droplets or perchlorate particles present in air within a work place, such as in a perchlorate manufacturing facility, may result in exposure to perchlorate. There are limited data on the availability of nonvolatile chemicals in shower droplets, although the low vapor pressure of perchlorate prevents it from coming out of solution and existing as a vapor. Furthermore, a study of the effects from inhaling airborne perchlorate particles in a manufacturing facility found no effect on thyroid function relative to other occupational workers (see Table 5.2).

Considering its soluble and persistent nature, the primary route by which humans may be exposed to perchlorate is ingestion of contaminated groundwater or surface water. Farmers may also use contaminated groundwater or surface water to irrigate food crops or water livestock. Therefore, indirect routes of human exposure could include ingestion of produce from farms irrigated with perchlorate-contaminated water or consumption of cow's milk containing perchlorate. These potential routes of exposure are discussed further below.

5.1.3.1 Exposure through Public Water Supplies

Since 1999, the U.S. EPA has required public water suppliers to monitor for perchlorate under the Unregulated Contaminant Monitoring Rule (UCMR) (40 CFR Parts 9, 141, 142). These monitoring data provide the U.S. EPA with information regarding the prevalence of perchlorate in water supplies so that the Agency can determine whether or not to regulate it. Table 5.3 summarizes the perchlorate data for potable groundwater and surface water, by U.S. EPA region, from data collected under the first round of the UCMR [30].

In 2004, the American Water Works Association reviewed perchlorate data from the U.S. EPA UCMR program and data from drinking water surveys by four states (Arizona, California, Massachusetts, and Texas). These data show that perchlorate has been detected at less than 12 μg/L in drinking water in 26 states and Puerto Rico [29]. The study reported that highest density of perchlorate detections occurred in southern California, in west central Texas, along the east coast between New Jersey and Long Island, and in Massachusetts. Perchlorate was not detected in drinking water in the northern Great Plains, the central and northern Rocky Mountains, Alaska or Hawaii. There was no reported difference in the rate of perchlorate occurrence between surface and groundwater used as drinking water sources.[1] Furthermore, detections of perchlorate in drinking water did not significantly correlate to known releases of perchlorate to the environment [6].

In California, more than 390 of the 5,512 sources for public water have contained perchlorate. California also has the greatest number of confirmed releases of perchlorate generally associated with rocket manufacturers or military facilities [31]. However, as indicated in Table 5.3, the majority of the detects in water samples obtained from the public water supplies as part of the UCMR program are less than the U.S. EPA's Drinking Water Equivalent Level of 24.5 μg/L. In other words, 95% of the total number of water samples with detectable levels of perchlorate contained perchlorate at concentrations less than 24.5 μg/L (619 of 647 total detections from Table 5.3). A survey of public water supplies in Massachusetts, including both surface and groundwater sources, found detectable levels of perchlorate in only nine of some 600 public water supplies [32]. In a survey of public water systems, private wells, irrigation wells, and USGS monitoring wells in 54 counties in west Texas and two counties in eastern New Mexico, perchlorate was present in 80% of the wells tested, with approximately 75% of the samples containing perchlorate concentrations below 4 μg/L and 25% of the samples equal to or greater than 4 μg/L [33].

Case Study: Kerr-McGee/PEPCON, Henderson, Nevada

One of the largest identified releases of perchlorate occurred at these facilities The release from Kerr-McGee operations has been linked to perchlorate-contam inated groundwater that discharged into the Las Vegas Wash and, subsequently into Lake Mead and the Colorado River. Lake Mead and the Colorado River are surface water sources to seven states in the United States (Arizona, New Mexico Colorado, Wyoming, Utah, Nevada, and California) and to Mexico. The lowe Colorado River is managed as a source of drinking water for roughly 15 to 2(million people in the states of California, Arizona and Nevada [34].

Surface water concentrations of perchlorate in the lower Colorado River belov the Hoover Dam have been detected at concentrations less than 4 μg/L since Jun 2004 [35]. In addition to being a source of drinking water, the water fron Colorado River provides enough water to irrigate 3.5 million acres of farmland fo commercial agriculture products [36].

---

[1] The authors of the study [6] noted the following caveats: the U.S. EPA UCMR database for small water supplies (those serving populations under 10,000 people) and surface water sources was largely incomplete and therefore under-represented; differences in analytical methods resulted in differences in detection rates, with those methods employing lower detection limits resulting in increased frequency of detections.

Table 5.3 Summary of UCMR 1 Survey of Perchlorate in Public Water Supplies (2001-2005) [30]

EPA Region	Frequency of Detection by Sample(2)			Number of Public Water Supplies(1) with Detected Perchlorate (>4 µg/L)		Frequency of Detection by Public Water Supply(3)	Range in Detected Concentrations (µg/L)		Average Concentration (µg/L)	Number of Samples Detected at Concentrations ranging from			
										4 to 6 µg/L	7 to 18 µg/L	19 to 24 µg/L	Greater than 25 µg/L
1	1	1,908	0.1%	237	1	0.4%	6	6	6	1	0	0	0
2	62	4,147	1.5%	377	18	4.8%	4	420	13	38	23	0	1[4]
3	13	2,049	0.6%	303	11	3.6%	8.3	37	15	6	4	2	2
4	46	5,308	0.9%	862	25	2.9%	8.3	220	25	22	17	1	6
5	14	3,028	0.5%	603	12	2.0%	4	32	10	8	4	0	2
6	24	3,140	0.8%	471	15	3.2%	4	32	12	6	14	2	2
7	2	1,132	0.2%	177	1	0.6%	4.9	7.2	6	1	1	0	0
8	0	1,262	0.0%	112	0	0.0%	<4	<4	<4	0	0	0	0
9	469	11,322	4.1%	501	72	14.0%	8.46	73.8	15	231	213	9	16
10	16	1,299	1.2%	165	5	3.0%	4	9	6	13	3	0	0
Total	647	34,595		3,808	160					326	279	14	28

Notes:

(1) Water samples include both surface water and groundwater sources for public water supplies [30].

(2) Frequency of detection represents the total number of water samples where perchlorate was detected divided by the total number of water samples analyzed for perchlorate.

(3) Frequency of detection represents the number of public water supplies where perchlorate was detected in at least one water sample divided by the number of public water supplies where perchlorate was analyzed in at least one water sample.

(4) A single sample contained detectable perchlorate greater than 240 µg/L.

5.1.3.2 Exposure through Food Supplies

Various studies have found perchlorate in plants used as food for both humans and animals. The U.S. Food and Drug Administration (FDA) conducted a preliminary market basket survey of perchlorate concentrations in commercially grown lettuce [37]. They found perchlorate at concentrations ranging from 0.5 to 129 µg/kg in 116 of 127 lettuce samples purchased in five states (Arizona, Texas, Florida, California, and New Jersey). Table 5.4 summarizes these data by state. Additional studies of food crops are being undertaken by the FDA.

Table 5.4 Summary of Perchlorate Concentrations (µg/Kg) in Lettuce by State [37]

State (where purchased)	Average Concentration	Standard Deviation	Range of Detected Concentrations		Number of Samples Detected	Number of Samples Analyzed
Arizona	14.8	20.6	3.7	129	35	35
Texas	7.6	0.83	6.8	8.5	3	3
Florida	21.8	33.8	0.5	71.6	3	4
California	8.1	8.4	0.5	52	67	78
New Jersey	8.1	6.9	1.2	21.6	8	8

Other researchers have detected perchlorate in lettuce (*Lactuca sativa* L.), tobacco (*Nicotiana tabacum*), soybeans (*Glycine max*), alfalfa (*Medicago sativa*), tomato (*Solanum lycopersicum*), and cucumber (*Cucumis sativus* L.) under laboratory conditions[2] [38] [39] [40]. Field studies of plants irrigated with water containing perchlorate have been conducted on a variety of agricultural plants consumed by humans, including lettuce (*Lactuca sativa*), winter wheat (*Triticum aestivum* L.) crops, alfalfa, watercress (*Nasturtium* spp.), chinaberry (*Melia azedarach*) and mulberry (*Morus alba*) trees, cucumber, cantaloupe (*Cucumis melo*), and tomatoes [39] [40] [41] [42]. Individually these studies include small sample sizes and have other limitations[3], but taken together, they indicate perchlorate uptake occurs during periods of active plant

---

[2] In general, laboratory studies are not designed to simulate actual field conditions but simply to determine if it is possible for perchlorate to be taken up by the plant of interest. Such laboratory studies typically maximize plant growth by maintaining optimum growth conditions (light, temperature, nutrients) and maximize potential perchlorate uptake by using sand as a growing medium and applying perchlorate-containing water daily. While laboratory studies represent a valuable tool to determine if perchlorate bioaccumulates in plants, the results need to be confirmed in field studies, where actual food crops are tested under ambient conditions normal for that particular plant/crop. The results from actual field studies of food crops represent a better indicator of potential exposure(s) to humans.

[3] Methodological problems include correlating perchlorate concentrations in irrigation water with measured concentrations in plants; controlling for potential confounders such as the affect of seasonal variations in precipitation; low sample numbers reflecting low power to demonstrate statistical difference should one exist based solely on perchlorate treatment versus natural variation; no control group; analytical methods [40] [42].

growth, with uptake and concentrations increasing with longer growing seasons. These studies also indicate that perchlorate accumulates primarily in leaf tissue, with lower concentrations associated with stems, pods, fruits, seeds, and roots, in decreasing order.

Agricultural animals may drink water containing perchlorate or consume grain from crops irrigated with perchlorate-contaminated water. In a study of beef cattle that ingested a mean concentration of 25 μg/L of perchlorate in water over a 14-week period, researchers found no significant change in the measurable perchlorate residues in blood plasma or edible tissue, nor was there any corresponding change in thyroid hormones between the exposed cattle and the control cattle [43]. The authors estimated that the heifers ingested 125 to 350 mg of perchlorate per day through drinking water from a spring-fed stream near the Naval Weapons Industrial Reserve Plant in McGregor, Texas. The water and plants adjacent to the stream contained measurable amounts of perchlorate [42], with water concentrations ranging from 20 to 60 μg/L. Furthermore, Batjoens (as cited by Cheng [43]) noted that even with prolonged perchlorate exposures (4 grams per day over 10 days), perchlorate may be measurable in cow's blood plasma but does not accumulate in tissue due to its short half-life. Thus, human exposure to perchlorate through consumption of meat from agricultural animals which have ingested perchlorate is unlikely to be a significant exposure pathway.

Researchers have also analyzed the concentration of perchlorate in cow's milk. In an initial FDA survey of cow's milk purchased from grocery stores across 14 states[4], perchlorate levels ranged from 3.16 to 11.3 μg/L in 101 out of 104 samples, with perchlorate not quantifiable in three samples [37]. The detected concentrations of perchlorate averaged 5.76 μg/L. Perchlorate has been identified in cow's milk by other published studies. Analysis of six samples from Utah found a maximum concentration of 6.22 μg/L [44]. Kirk et al. [45] [46] detected perchlorate at concentrations ranging from 0.47 μg/L to 11.0 μg/L in 54 milk samples purchased in 12 states. Table 5.5 summarizes the available data on perchlorate in cow's milk by state from these studies.

The California Department of Food and Agriculture measured perchlorate levels ranging from 1.5 to 10.6 μg/L in California milk, and the Environmental Working Group (EWG) reported levels up to 3.6 μg/L in June 2004 (as cited by the FDA [47]). The results from these efforts are within the ranges reported in other studies [37] [45] [46].

A few studies have evaluated perchlorate concentrations in human breast milk. Kirk et al. [46] measured concentrations of perchlorate ranging from 1.4 to 92.2 μg/L in 36 samples of human milk from 18 states, as shown in Table 5.6. This phenomenon has been also been observed in laboratory animals. A study on rats found perchlorate at higher concentrations in the milk than in the blood serum of previously dosed animals (Clewell et al. as cited in [48]).

The FDA is further evaluating perchlorate levels in various foods, including milk, bottled water, and produce such as lettuce, spinach, tomatoes, carrots, and cantaloupe. Perchlorate was found in lettuce, bottled water, and milk samples purchased at retail stores from various regions of the country. The FDA considers

---

[4]The milk products included unpasteurized milk, pasteurized whole milk, 2% milk, 1% milk, chocolate 1% milk, fat free milk, and organic whole milk.

Table 5.5 Summary by State of Perchlorate Concentrations (µg/L) in Cow's Milk

State Where Milk Was Purchased	Average Concentration	Range in Detected Concentrations		Number of Samples Detected	Number of Samples Analyzed	Reference
Arizona	6.1	3.2	10.4	19	20	[37]
Arizona	1.1	0.8	1.7	4	4	[45] [46]
Arkansas	1.2	0.9	1.6	7	7	[45] [46]
California	5.7	3.4	9.9	38	38	[37]
California	3.5	0.8	11.0	10	10	[45] [46]
Florida	1.5	1.1	2.0	3	3	[45] [46]
Georgia	8.7	NQ	9.1	2	3	[37]
Hawaii	2.3	1.0	5.7	4	4	[45] [46]
Kansas	10.4	10.4	10.4	1	1	[37]
Kansas	1.5	0.6	2.6	5	5	[45] [46]
Louisiana	4.4	3.2	7.0	5	5	[37]
Maine	0.5	0.4	0.6	3	4	[45] [46]
Maryland	6.3	3.4	11.3	7	7	[37]
Missouri	6.3	4.6	7.9	4	4	[37]
New Hampshire	4.7	4.7	4.7	1	1	[45] [46]
New Jersey	4.1	3.4	4.8	5	5	[37]
New Mexico	1.2	0.7	2.0	3	3	[45] [46]
New York	4.4	4.4	4.4	1	1	[45] [46]
North Carolina	5.1	5.1	5.1	1	1	[37]
Pennsylvania	6.7	NQ	7.2	3	4	[37]
Pennsylvania	1.6	0.8	2.8	5	5	[45] [46]
South Carolina	6.3	6.3	6.3	1	1	[37]
Texas	5.6	3.8	7.2	5	5	[37]
Texas	4.5	1.8	6.4	7	7	[45] [46]
Utah	NQ	2.9	6.2	6	6	[44]
Virginia	7.2	6.8	7.6	3	3	[37]
Washington	5.5	3.2	7.7	7	7	[37]
Maximum	10.4	10.4	11.0			
Minimum	0.5	0.4	0.6			

Notes

(1) NQ = non-quantifiable

(2) The estimated limit of quantitation (LOQ) is 3.0 µg/L for milk in the data from reference [37] and 0.4 µg/L (IC) and 1 µg/L (IC–MS) in data reported in [46].

(3) Data are presented to two significant figures, which may differ from that published in cited study.

the levels detected to be preliminary and not reflective of the distribution of perchlorate within the U.S. food supply because of limitations in the number of food categories, specific batches or production lots, and brands. The Centers for Disease Control (CDC), in collaboration with the FDA, are completing additional studies to determine the scope and public health implications of perchlorate levels in various foods [37].

Table 5.6 Summary of Perchlorate Concentrations (µg/L) in
Human Breast Milk by State [46]

State	Range of Detected Concentrations		Number Samples
California	1.4	20.7	5 (average 6.6)
Connecticut	1.4	1.4	1
Florida	1.9	1.9	1
Georgia	1.9	2.6	3 (average 2.1)
Hawaii	1.6	1.6	1
Maryland	4.1	4.1	1
Maine	2.3	2.3	1
Michigan	4.7	4.7	1
Missouri	31.6	31.6	1
North Carolina	3.8	3.8	1
Nebraska	31.5	31.5	1
New Jersey	3.2	92.2	3 (average 48.7)
New Mexico	37.6	37.6	1
New York	3.8	3.8	1
Texas	1.4	12.7	10 (average 5.32)
Virginia	3.3	3.3	1
Washington	2.3	2.3	1
West Virginia	2.3	2.3	1

5.1.4   Risk Characterization

The studies described above have shown that humans may be exposed to perchlorate through drinking water or ingesting certain foods. The dose-response assessment showed that perchlorate does not persist in the human body and that the only known effect of low-level perchlorate exposure is the reversible inhibition of iodide uptake by the thyroid gland. So what are the risks from environmental exposure to perchlorate?

The NRC report [5] concluded that the best point-of-departure to assess a safe intake of perchlorate was a clinical study conducted by Greer et al. [49], who dosed healthy adult volunteers and used a very sensitive indicator, the uptake of radioactive iodide, to determine what dose would not interfere with the normal function of the thyroid gland. The lowest dose that showed no inhibition of the uptake of radioactive iodide was 0.007 mg/kg•day. This dose, which is truly a No Observed Effects Level (NOEL), was then divided by a safety factor of 10 to protect sensitive human populations, including pregnant women, fetuses, and newborn babies. The U.S. EPA accepted this value of 0.0007 mg/kg•day as the basis for their Reference Dose (RfD) [10], which is the amount of perchlorate that can be ingested daily without any concern about causing an adverse effect. The U.S. EPA's Integrated Risk Information System (IRIS) database states that [10]:

Iodide uptake inhibition is a key biochemical event that precedes all potential thyroid-mediated effects of perchlorate exposure. Because iodide uptake inhibition is not an adverse effect but a biochemical change, this is a No Observed Effect Level (NOEL). The use of a NOEL differs from the traditional approach to deriving an RfD, which bases the critical effect on an adverse outcome. Using a nonadverse effect that is upstream of the adverse effect is a more conservative and health-protective approach to perchlorate hazard assessment.

The RfD chosen for the protection of human health, being based on a no-observed-effect-level that is, in turn, divided by an "intraspecies" uncertainty factor of 10, indicates a safe, conservative dose for any human receptor, including sensitive populations like pregnant women and children. (Chapter 6 provides further information on regulatory limits on perchlorate in drinking water.)

Derivation of Reference Dose

The U.S. EPA develops an RfD [mg/kg•day] to represent the safe <u>daily</u> dose of a noncarcinogenic chemical for any human receptor (including pregnant women and children), conservatively based on testing data and incorporating uncertainty factors. In the case of perchlorate, the No Observed Effect Level (NOEL) was the lowest human dose that would not interfere with the uptake of iodide by the thyroid gland. This dose was 0.007 mg/kg•day. To obtain the RfD of 0.0007 mg/kg•day, the U.S. EPA divided that dose by an uncertainty factor of 10 to protect sensitive people [10].

Based on current information regarding the magnitude of likely exposures (e.g., drinking water and food), all of the currently available occupational and epidemiological data and the considerable weight-of-evidence on the toxicity and hazard of perchlorate in the literature, it is reasonable to conclude that perchlorate does not pose a significant hazard to the vast majority of people living in the United States.

5.2 Overview of Ecological Risk Assessment

Ecological risk assessment is more complex and, because of a lack of historical or biological data, generally less certain than methods used for human health risk assessment. Rather than considering effects on a population of a single species - that is, humans - ecological risk assessors must evaluate the risks to ecosystems containing perhaps thousands of interdependent species. These animals range from terrestrial mammals and birds to fish and aquatic invertebrates. Unlike human health risk assessment, ecological risk assessment looks at potential effects on populations and higher assemblages, such as communities and ecosystems. The process parallels human health risk assessment in its consideration of basic exposure routes and potential health effects. One clear difference is a lack of cancer endpoints for animals due to the fact that short life spans generally preclude the time required for initiation and induction of tumors.

5.2.1   Ecological Exposures

When addressing environmental hazard(s), ecological risk assessors typically address aquatic and terrestrial ecosystems separately because the distribution and fate of contaminants is markedly different between them. Both systems, of course, are rarely mutually exclusive at actual sites. The subsections below discuss exposures in both types of ecosystems.

5.2.1.1   Aquatic Exposure

This section principally addresses concentrations of perchlorate in surface water and sediments and the exposures of aquatic organisms to these concentrations. According to U.S. EPA [31], at least 25 states have confirmed releases of perchlorate. In general, because perchlorate is very soluble in water, perchlorate concentrations are highest in surface water and sediments near specific releases, but decline rapidly with increasing distance from the source. Perchlorate apparently also degrades rapidly in anaerobic sediments, provided sufficient substrate availability and lower concentrations of competing anions such as sulfate and nitrate [81]. The case studies which follow illustrate aquatic exposures resulting from known perchlorate releases.

Case Study: Kerr-McGee/PEPCON, Henderson, Nevada
Releases of perchlorate from the former Kerr-McGee plant migrated through groundwater to the Las Vegas Wash, then through surface water flow to Lake Mead. Surface water samples from the Las Vegas Wash have contained perchlorate at concentrations as high as 130 mg/L and sediment at concentrations as high as 56 mg/kg. Perchlorate was also detected in fish and aquatic vegetation at concentrations as high as 44.3 mg/kg and 176 mg/kg, respectively. However, these concentrations rapidly decreased with distance from the source area as a result of dilution of the perchlorate plume by the waters of the Las Vegas Wash and Lake Mead. Downstream near Lake Mead, researchers found perchlorate in surface water at concentrations as high as 0.068 mg/L and in sediment at concentrations as high as 0.081 mg/kg. Perchlorate was not detected in fish, amphibians, aquatic invertebrates, or aquatic vegetation at downgradient locations [50].
Subsequently, Smith et al. [51] collected surface water and unspecified aquatic vegetation samples over an approximately two mile section of the Las Vegas Wash. The mean concentrations of perchlorate in surface water ranged from 0.15 to 1.04 mg/L. In aquatic broadleaf plants, the mean perchlorate concentration ranged from 6.6 to 545 mg/kg, while the levels in aquatic grasses ranged from 3.1 to 10.5 mg/kg. The researchers postulated that the greater leaf surface area of broad leaf plants may result in higher transpiration rates and therefore account for the higher perchlorate concentrations.

Case Study: Naval Surface Warfare Center, Indian Head, Maryland
Parsons [50] sampled six locations downstream from one area of the site, the Low Vulnerability Ordnance Area (LVOA), in the Town Gut Marsh System. Surface water samples at five of the six locations contained perchlorate at concentrations ranging from 0.004 to 0.025 mg/L. Sediments at the three most upstream locations contained detectable perchlorate, at concentrations ranging from 0.004 to 0.318 mg/kg. Perchlorate was not detected in the three locations furthest downstream. Invertebrates collected at two downstream locations did contain detectable concentrations of perchlorate. Of the amphibians collected at four downstream locations, only one - at the most upstream location - contained detectable perchlorate, at 0.465 mg/kg. Fish from the four downstream locations did not contain detectable concentrations of perchlorate.

Additional work has been done to characterize environmental concentrations of perchlorate at the Naval Weapons Industrial Reserve Plant (NWIRP) in McGregor, Texas. The studies described below examined the levels of perchlorate in surface water, sediments, and fish.

Theodorakis et al. [52] found that perchlorate concentrations in stream were highly variable over seasons, ranging from undetectable to approximately 0.150 mg/L. Some concentrations exceeded the chronic screening benchmark (Section 5.2.2) and were affecting thyroid homeostasis in fish. Theodorakis et al. expressed uncertainty whether these levels of perchlorate and/or thyroid impacts could lead to effects on growth, development, or survival in field populations.

Tan et al. [42] looked at four sample locations at the NWIRP where sediments had been either intermittently or continuously exposed to perchlorate in surface water. The intermittently exposed stream segments were downstream of the continuously exposed stream segments, and Tan et al. [42] suggested that some of the lower perchlorate concentrations in the intermittently exposed streams were attributable to bacterial activity in wetland habitats. The team analyzed perchlorate concentrations in surface water, smartweed, and watercress. Table 5.7 summarizes the results.

Table 5.7 Results of Perchlorate Sampling at NWIRP [42]

Sample (concentration units)	Perchlorate Concentrations	
	Continuously Exposed Sediments (upstream)	Intermittently Exposed Sediments (downstream)
Surface water (mg/L)	0.006-0.536	ND-0.123
Smartweed (mg/kg)	5.85-61.6	ND-10.5
Watercress (mg/kg)	ND-13.9	ND-7.9

Subsequently, Tan et al. [53] examined the potential for anaerobic degradation of perchlorate at these four sample locations. As described in more detail in Chapter 4, their laboratory mesocosm experiments showed that bacteria in anaerobic sediments can degrade perchlorate, although high concentrations of nitrate may affect degradation rates.

Tan et al. [54] then looked at the temporal and spatial variation of perchlorate in streambed sediments at NWIRP. As described in Chapter 4, the researchers suggested that plant growth may contribute organic substrate for perchlorate degradation, and plant uptake may also have contributed to perchlorate removal from sediments. Overall, Tan et al. [54] concluded that the sediments were able to degrade perchlorate concentrations within the top 30 centimeters.

Theodorakis et al. [52] looked at perchlorate concentrations in fish at NWIRP. They examined both the heads and the fillets, examining the former "because they may be a more sensitive indicator of perchlorate accumulation in fish than the fillets." In those streams where perchlorate was detected in the surface water, it was also detected (although infrequently) in fish. Theodorakis et al. [52] reported a general pattern of more frequent detection of perchlorate in fish captured closer to NWIRP.

As these case studies show, perchlorate generally does not accumulate or persist in sediments due to its solubility and microbial degradation under the proper anaerobic conditions. Therefore, if the source of perchlorate is surface water, then sediment contamination may not be as important an issue as it might be if the perchlorate source is groundwater. Uncontrolled groundwater discharge can represent a steady, continuing source of perchlorate to sediments. Further, analyses of specimens at varying levels in the food chain seem to indicate that perchlorate does not have a high potential to bioaccumulate.

5.2.1.2   Terrestrial Exposure

Because perchlorate is highly water soluble, most of the focus with regard to environmental problems has been on groundwater contamination and subsequent off-site migration in groundwater or surface water. As such, information addressing the exposure of terrestrial wildlife to perchlorate is limited and, of that available, most is derived from controlled laboratory studies.

Very high concentrations of perchlorate (percent levels) are rare in the general terrestrial environment, occurring primarily at manufacturing facilities or at sites that tested or burned formulations or explosives containing high concentrations (e.g., up to 80% by weight) of the oxidizer. Examples of these types of facilities include certain U.S. military reservations and/or target ranges, test firing and/or open burning ranges and rocket engine developers.

Soil at other locations may contain trace levels (μg/kg) of perchlorate. The best example at the local level would be areas that were routinely used as an ignition platform for fireworks displays. Perchlorate is also used as an oxidizer in roadside flares, so very limited soil exposure may occur in areas where flares were used (or ditches that may have received stormwater runoff from these areas). Finally, perchlorate is found naturally in some types of imported inorganic fertilizers [32]. Although the use of these types of fertilizers has been curtailed because of the presence of perchlorate, areas that previously received these types of soil amendments may continue to present ecological receptors with low levels of perchlorate in soil.

While terrestrial animals may drink surface water containing perchlorate, most remedial investigations reveal that this route is generally not a significant exposure pathway for terrestrial organisms. Exceptions may occur at a handful of hazardous waste or military sites where surface waters may be actively fed by contaminated groundwater which contains concentrations of perchlorate in the mg/L range. Thus, this discussion of ecological exposures principally addresses concentrations of perchlorate that might be encountered in a terrestrial setting (e.g., military reservation or test range), principally from soil and plants or forage.

Perchlorate in Soil

Surface soil can be the first medium to receive a "dose" of perchlorate as a contaminant, typically from an explosive device or the open burning of explosives or munitions. Because perchlorate is very water soluble, the concentrations in soil will markedly decrease with time as rainfall infiltrates the soil. Perchlorate concentrations in arid regions, which receive very little rainfall, would persist in shallow soil for a longer period of time.

The U.S. EPA drafted a *Screening Level Ecological Risk Assessment for Perchlorate* [8], which cites a concentration range "from less than 1 to 1,470 mg/kg" of perchlorate in soil sampled from various sites. Although a careful review of the literature shows a few sites where the maximum value of perchlorate in soil exceeds this concentration range, it may be misleading to focus on the highest detected concentrations because soil contamination at many sites tends to spatially distribute in "clusters" or "hot spots," for example those located in and around targets used at military ranges. The frequency distribution of perchlorate concentrations, like most chemicals in soil at hazardous waste sites, is therefore log normal and heavily skewed to the left, meaning the majority of soil analyses would be observed below the detection limit. As a consequence, most terrestrial animals inhabiting a site, due to the selective nature of foraging activity, are anticipated to be exposed to, on average, lower soil levels than the maximum values which are sometimes cited.

Case Study: MMR, Cape Cod, Massachusetts
Less than 10% of the soil samples (216 out of 2,500) at MMR contained detectable perchlorate (range was 0.0012 to 27 mg/kg, with a median and mean concentration of 0.0021 and 0.252 mg/kg, respectively) [55]. A careful review of the literature shows that most soil samples at hazardous waste sites (that have a known history of exposure to perchlorate) are less than 20 mg/kg [9].

Alsop et al. [60] studied perchlorate uptake by plants and small mammals in one area at MMR. Soil in the study area contained perchlorate concentrations that ranged from non-detect to a maximum concentration of 27 µg/kg. They reported that perchlorate was taken up by plants, and the perchlorate concentrations ranged between non-detect and 140 µg/kg. Perchlorate was not detected in earthworms exposed to perchlorate containing soil. Perchlorate was not detected in mice or voles collected from the study area, but it was detected in a single shrew sample at 140 µg/kg.

Perchlorate in Plants

Plants can be a significant source of food for terrestrial organisms, especially strict herbivores. Laboratory studies have clearly demonstrated that perchlorate, being a dissociable salt, is readily taken up by most types of plants [8]. The rate of translocation from the soil to the plant, however, varies markedly with the type of plant tested, the concentration of perchlorate in the soil, and, in some cases, the season the plant matter was sampled.

Perchlorate appears to concentrate in plants via a simple "salting out" mechanism. The perchlorate ion is taken up into the plant from soil pore water, translocated to the leaves and/or distal portions of the plant and, through evapotranspiration, is concentrated in the leaves (although seeds typically have a low concentration of perchlorate). This type of mechanism is important from the standpoint of both human and ecological exposures.

Bioconcentration factors, which are simply the ratio of the concentration of perchlorate in the plant to the concentration in the soil, generally range from 2 to 200 [8] (a few citations report higher bioconcentration factors but those studies included more toward aquatic plants or hydroponic testing). Some plants, such as kelp, appear to bioconcentrate perchlorate from natural sources. Others, like tobacco fertilized with Chilean salt peter, appear to accumulate the salt to a much higher degree than other plants tested [56]. The U.S. EPA Screening Ecological Risk Assessment adopted "a simple, conservative, screening-level" bioconcentration factor of 100 for use in estimating plant concentrations from available soil data [8]. Although soil-to-plant bioconcentration factors tend to agree from study to study, it is important to note that some studies report concentrations as dry weight but others will report wet weight. Because of vast regional differences in both soil chemistry and weather, it is best to consider site-specific sampling of vegetation rather than relying on a "default" bioconcentration factor.

Wild or inedible plants have primarily been sampled from areas with naturally occurring elevated levels of perchlorate (e.g., in the arid areas of the southwestern U.S.), or impacted military sites. Therefore the analytical results reported are most likely biased high rather than representative of a broad range of conditions. With this caveat in mind, the available literature cites concentrations of perchlorate in samples of wild plants (inclusive of grasses, shrubs, succulents and trees) ranging from below the limit of detection (ND) to 5,500 mg/kg [40] [51] [57] [58] [59]. Most values measured in vegetation, however, are in the low-to-mid parts per billion range (trees, ND-220 μg/kg; succulent cactus, 66-3,200 μg/kg; desert scrub, 16-900 μg/kg; winter wheat, 720-8,600 μg/kg fresh weight (FW); garden vegetables, 40-1,650 μg/kg FW).

Market basket surveys of grocery produce have shown that lettuce, which has an appreciably high water content, may contain significant levels of perchlorate. This may result from the fact that much of the U.S. lettuce crop is grown on California farms that are irrigated with water from the Colorado River (containing an average of 6 μg/L of perchlorate) [8] [36]. The frequency of detection and concentration of perchlorate in the lettuce samples varies according to the study, perhaps in part because some investigators included the outer leaves (which are prone to field soil contamination), whereas others sampled only the "edible core." Section 5.1.3 provides additional information on these studies.

5.2.2   Ecological Effects

Studies have shown that surface waters, and to a lesser extent sediments, may contain perchlorate to which fish, birds, and mammals may be exposed. Similarly, plants, which birds and mammals may use for food, may contain perchlorate. This section reviews the ecotoxicology of perchlorate, including available dose-response information, to evaluate the potential effects of perchlorate exposure to aquatic and terrestrial organisms.

5.2.2.1 Effects on Aquatic Receptors

Studies have shown that exposure to perchlorate at high concentrations can cause mortality in fish, although some of the results have been confounded by the potential toxicity of the cation (e.g., ammonium or sodium) rather than toxicity of the perchlorate anion *per se* [8] [63]. It has also been noted that exposure to perchlorate may delay metamorphosis in frogs, although the effect appears to be reversible. These findings are described below.

In developing Ambient Water Quality Criteria (AWQC) for perchlorate, the U.S. EPA [63] considered the data shown in Table 5.8. (Section 6.1.3 describes the derivation of AWQC for perchlorate.) The studies provided 96-hour LC50 data, or in other words the concentrations lethal to 50% of the test organisms after a 96-hour exposure period. When more than one freshwater LC50 is available for a species within a genus, the geometric mean of all the data are calculated to represent the genus mean acute value (GMAV).

Table 5.8 Studies Evaluating the Effects of Perchlorate
on Aquatic Organisms [63]

Type	Species	Acute LC50 Values (mg/L)	GMAV (mg/L)	Rank
Invertebrate	*Ceriodaphnia dubia*	77.8, 66	71.7	1
Invertebrate	*Daphnia magna*	490	490	2
Fish-Cyprinidae	*Pimephales promelas*	1.655, 614, >490	793	3
Invertebrate	*Hyalella azteca*	>1,000	>1,000	4
Fish-Centrarchidae	*Lepomis macrochirus*	1,470	1,470	5
Fish-Salmonidae	*Oncorhynchus mykiss*	2,010	2,010	6
Invertebrate	*Lumbriculus variegatus*	3,710	3,710	7
Amphibian	*Rana clamitans*	5,500	5,500	8
Invertebrate	*Corbicula fluminea*	6,680	6,680	9
Invertebrate	*Chironomus tentans*	8,140	8,140	10

Some questions arose regarding these data. For example, U.S. EPA [63] notes that, in the *Pimephales promelas* toxicity test that resulted in an LC_{50} of 614 mg/L, sodium chloride at the same sodium concentration caused toxicity. The reported results may reflect sodium toxicity rather than perchlorate toxicity.

The U.S. EPA also looked at the potential for perchlorate exposure to disrupt endocrine function at particular points in the frog life cycle. U.S. EPA [64] reviewed data from LHAAP (see case study) and unpublished work by Tietge and Deitz that evaluated the effect of perchlorate on the development and metamorphosis of the African clawed frog (*Xenopus laevis*). This work indicated a no effect concentration of 0.062 mg/L and a low effect concentration of 0.25 mg/L. Tietge et al. [65] later published this work.

Case Study: Longhorn Army Ammunition Plant (LHAAP), Karnack, Texas
Since 1998, treated groundwater has been pumped to a wastewater holding pond for storage and then discharged to Harrison Bayou when there is flow in the bayou. Treatment did not remove perchlorate until March 2001. Researchers have tested perchlorate concentrations in a variety of environmental media, as summarized below.

Sample (concentration units)	Perchlorate Concentrations September 2000 [50]		Perchlorate Concentrations November 1999 [51]	
	Wastewater Holding Pond	Harrison Bayou	Wastewater Holding Pond	Harrison Bayou
Surface water (mg/L)	Up to 3.8	ND	30.8-31.4	ND-0.004
Sediment (mg/kg)	Up to 0.286	ND	12.2-35.6	ND
Aquatic vegetation (mg/kg)	Up to 15.7	ND	Bulrushes: AWL 5.98-9.49 BWL 2.42-7.46 R 0.56-1.13	NA
Aquatic invertebrates (mg/kg)	NA	ND	NA	NA
Aquatic insects (mg/kg)	Up to 0.996	ND	0.81-2.04	NA
Amphibians (mg/kg)	Up to 1.66	ND	Bullfrog tadpoles 1.28-2.57	Frogs ND-0.153
Fish (mg/kg)	NA	ND	NA	0.077-0.206

ND - not detected; NA - not analyzed; AWL - above water line; BWL - below water line; R - roots

As these data show, surface water in the Wastewater Holding Pond has contained perchlorate at concentrations above all three of the U.S. EPA's unpromulgated criteria: acute AWQC (22.3 mg/L), chronic AWQC (10.3 mg/L), and the interim chronic benchmark (0.12 mg/L). However, concentrations downstream in Harrison Bayou did not exceed these criteria in these studies.

U.S. EPA [63] reviewed work by Goleman et al. [66] that looked at the effects of perchlorate on the metamorphosis of the clawed frog (*Xenopus laevis*). The perchlorate concentrations detected (0.059 to 14.4 mg/L in water) bracketed the concentrations found at LHAAP. The tests showed that perchlorate concentrations as low as 0.018 mg/L inhibited forelimb emergence and tail resorption but did not affect growth in a 70-day exposure period. Further work by Goleman et al. [67] indicated that the effects seen on forelimb emergence and tail resorption at 70 days were reversible in a 28-day recovery period in control solutions. They interpreted this result as consistent with the mechanism of perchlorate action being competitive inhibition of iodide uptake by the thyroid.

U.S. EPA [63] also reported on research by Carr et al. [68] on native frogs at LHAAP, including bullfrogs and chorus frogs. That research suggested that concentrations of perchlorate in freshwater on the order of 2.0 to 10 mg/L affected metamorphosis in frogs. However, Theodorakis et al., as cited by U.S. EPA [63], found no effects of these waters (0.1-5 mg/L) on native bullfrogs or *Xenopus laevis* metamorphosis. The U.S. EPA [63] noted that high iodide concentrations in these waters may have affected the study results.

One study at the LHAAP [69] examined the effect of localized perchlorate contamination on raccoons. Previous investigations at this site have shown relatively high concentrations of perchlorate (0.5-> 5,000 mg/kg) in vegetation (especially near contamination sources) but relatively low concentrations of perchlorate in wildlife (ND-< 2.3 mg/kg). A comparison of raccoons from uncontaminated vs. contaminated areas did not reveal any differences in plasma concentrations of any of the thyroid hormones or thyroid-stimulating hormone. The authors [69] concluded that "Analysis of perchlorate in common food items present at the LHAAP indicated that there was potential for exposure among raccoons. However, perchlorate does not appear to be a significant risk to raccoons at the LHAAP based on chemical analysis of plasma and quantitation of thyroid hormone concentrations."

5.2.2.2 Effects on Terrestrial Receptors

Assessment of the terrestrial effects of perchlorate has principally been limited to laboratory studies where the dose of the pure salt (typically ammonium perchlorate) can be evaluated against the response observed following controlled application to a plant or animal. Additionally, in order to observe concentrations that will have a No Observed Adverse Effect Level (NOAEL) and a Lowest Observed Adverse Effect Level (LOAEL), it is typical to employ dose ranges that may include concentrations over three orders of magnitude. Thus, many researchers acknowledge that the concentrations tested are conservative in the sense that concentrations studied in the laboratory may not exist in all but the most contaminated environments.

The ecotoxicology of perchlorate is still very much in its infancy. Scientists generally agree that the only mechanism by which perchlorate may manifest an adverse effect on an organism is by inhibiting the transport of iodide into the thyroid gland (through the competitive, but *reversible,* blocking of the sodium iodide symporter, or NIS). Therefore, ecological concerns following a release of perchlorate focus on animals that have a thyroid gland, such as birds and mammals. Additionally, since the thyroid gland regulates basic metabolic processes and a wide array of other essential tissue functions, more focus is now being placed on the secondary effect that perchlorate exposure may have on the developing fetus and newborn.

The U.S. EPA reviewed most of the research studies addressing the fate, transport and effect of perchlorate on terrestrial organisms in a draft toxicological profile entitled *Perchlorate Environmental Contamination: Toxicological Review and Risk Characterization* [8] [63]. As of early 2006, the draft document has received extensive comments from both public and private parties. It is not known when a final document will be issued. The known effects of perchlorate on terrestrial plants, invertebrates, and vertebrates are recapped below (Table 5.9), but the reader is encouraged to consult these comprehensive documents [8] [9] [63] should more detailed information be required.

Plants

Researchers have evaluated two types of effects of perchlorate on plants: vegetative stress, or inhibition of plant growth, and concentration of perchlorate in plants that may serve as a food source. Each of these types of effects is discussed below.

Plants usually serve only as crude indicators of soil contamination. To induce vegetative stress, soil concentrations of the vast majority of environmental contaminants generally have to be very high - that is, in the low percent range - to stunt or inhibit the growth of vegetation. In the case of perchlorate, there is no evidence that trace (μg/kg) soil concentrations will stress or inhibit the growth of plants. Researchers have used slightly higher concentrations (mg/kg) to evaluate the concentrations of perchlorate in soil or sand that will inhibit the growth of lettuce or cucumber seeds. Table 5.9 presents data from these studies. The estimated NOAEL for cucumber seed germination is approximately 10 mg/kg, while the concentration considered a LOAEL for lettuce is 10 times lower (1 mg/kg) [70] [71]. The U.S. EPA then derived a safe level in soil of 0.1 mg/kg by applying an uncertainty factor of 10 "to account for interspecies variance" which is expected to be protective of all plants. This screening value is highly conservative because the perchlorate applied was completely dissolved in irrigation water.

Most of the laboratory studies to date have used clean sand or soil and then irrigated the plants with water containing nominal (and widely varying) concentrations of perchlorate. It is therefore difficult to know how actual perchlorate residuals in soil may affect wild plants. After long periods of weathering, which allows perchlorate to travel deeper into the soil (or exchange with less tightly bound anions), there is less perchlorate in the surface soil and it is less available for plant uptake or animal exposure.

Perchlorate in plants is currently more of a concern from the standpoint of whether bioconcentration, via an evaporative salting out effect, can occur to such a degree that animals grazing on vegetation containing perchlorate salt may be at risk.

Soil Invertebrates

The few studies that have tested perchlorate toxicity to soil invertebrates (Table 5.9)

Table 5.9 Studies Evaluating the Effects of Perchlorate on Terrestrial Organisms Following Varying Levels of Exposure

| Species | Exposure: | | | | Endpoint | Estimated "No Effect" Concentration | Estimated "Safe" Exposure Concentration | Ref. |
	Medium	Location	Duration	Concentrations				
Lettuce Seed (Lactuca sativa)	Sand and Soil	Laboratory	28 Day	10-2,400 mg/kg	Germination	10 mg/kg	1 mg/kg[a]	[70]
Cucumber Seed (Cucumis sativus)	Sand	Laboratory	10 Day	0.01-10 mg/L (0.1-1 mg/kg)	Germination	10 mg/L (1 mg/kg)	10 mg/L (1 mg/kg)	[71]
Earthworm (Eisenia foetida)	Soil	Laboratory	14 Day	2,000-5,000 mg/kg	LC$_{50}$	N/A	1 mg/kg[b]	[70]
Rat (Rattus norvegicus)	Water	Laboratory	14 Day	0-30 mg/kg·day	Brain Morphometry	<0.01 mg/kg·day[c]	0.01 mg/kg (ww, plant tissue)	[72]
Deer Mouse (Peromyscus maniculatus)	Water	Laboratory	Gestation & 21 Post-Partum	0-117.5 mg/L	Thyroid Hormone Titers	<0.12 µg/L	<0.12 µg/L (drinking water)	[73]
Prairie Vole (Microtus ochrogaster)	Water	Laboratory	51 Days	0-10 mg/kg·day	Thyroid Hormone Titers	10 mg/kg·day	1 mg/kg·day	[74]

Table 5.9 Studies Evaluating the Effects of Perchlorate on Terrestrial Organisms Following Varying Levels of Exposure, continued

| Species | Exposure: | | | | Endpoint | Estimated "No Effect" Concentration | Estimated "Safe" Exposure Concentration | Ref. |
	Medium	Location	Duration	Concentrations				
Bobwhite Quail (Colinus virginianus)	Water	Laboratory	30 Days	0-1 mM (0-117.5 mg/L)	Toxicity; Thyroid Function; Egg Production	0.1 mM (117.5 µg/L)	0.1 mM (117.5 µg/L)	[75]
Mallard Duck (Anas platyrhynchos)	Water	Laboratory	14 Days	.025-250 mg/L	Thyroid Hormone Titers	1.0 mg/L	1.0 mg/L	[76]
Raccoon (Procyon lotor)	Food, Water, Soil	Field (LHAAP)	Lifetime	ND-5,500 mg/kg (food); ND-31 mg/L (water)	Thyroid Hormone Titers	No Apparent Effect Due to On-site Concentrations	<5,500 mg/kg food; <31 mg/L drinking water	[69]

[a] Derived by dividing NOAEL of 1 mg/kg by an interspecies uncertainty factor of 10.

[b] Derived by dividing LC50 of 4,445 by an interspecies variance factor of 242 and an assumed acute/chronic ratio of 18.

[c] Effects seen in brain morphometry of offspring, though results are highly controversial; safe dose estimated for "plant tissue" (wet weight).

utilized earthworms (*Eisenia foetida*) and dosed the test organisms with perchlorate via spiked irrigation water rather than spiking the soil prior to the test. The U.S. EPA cites a value of 4,450 mg/kg as the lowest LC_{50} calculated from the earthworm test [8]. The Agency then uses water quality criteria development methodology (which are not rigorously applicable to development of protective soil screening standards) to apply an interspecies uncertainty factor of 242 and an acute-to-chronic ratio of 18 to obtain a safe soil screening concentration of 1 mg/kg for soil-dwelling macroinvertebrates. They justify this three order of magnitude safety factor based on the limited data set. It is apparent from the results of the acute study that chronic soil values would most likely be well above this estimated safe concentration. The 1 mg/kg screening value is therefore a very conservative estimate of what might be considered a safe soil perchlorate concentration for invertebrates.

Herbivores

The U.S. EPA screening ecological risk assessment relies on data from laboratory experiments with experimental rats [77] [78] and deer mice [73]. Table 5.9 summarizes the highlights of both studies, but also includes a more recent study conducted on prairie voles. It appears that the LOAEL for laboratory rats from the Argus study [77] [78] was 0.01 mg/kg•day. This value is close to the LOAEL for deer mice, which, assuming a water ingestion rate of 0.34 g/g•day, was estimated to be 0.04 mg/kg•day. Prairie voles, when tested using slightly different toxicity endpoints, appeared to be less sensitive to perchlorate ingestion with a LOAEL of approximately 10 mg/kg•day. In rats, the U.S. EPA applied an additional interspecies uncertainty factor of 10 to the laboratory data to estimate a safe intake rate of 0.001 mg/kg•day [8], equivalent to a wet weight plant tissue concentration of 0.01 mg/kg.

From the results of these laboratory bioassays, the U.S. EPA suggests that herbivores such as mice, moles, and voles may be the most sensitive population at a site where perchlorate has been released to the general environment. One must consider, however, that these U.S. EPA estimates of safe intake rates are screening values that have incorporated worst case exposure scenarios and are therefore conservative benchmarks for the protection of wildlife.

Birds

Researchers have evaluated the effects of perchlorate exposure on bobwhite quail, both adults and chicks, and mallard ducklings, as summarized in Table 5.9. These studies have focused on exposure via drinking water.

An evaluation of adult bobwhite quail that were dosed with perchlorate via drinking water [75] showed that only the high dose level of 1 mM perchlorate (0.12 mg/L) affected the histopathology of the thyroid gland. Birds affected in the high dose group had a significant reduction in the colloidal area and a significant increase in the height of the thyroid follicular cells. All three dose groups showed no effect of perchlorate on the production of eggs, nor on the body or organ weights of the birds.

Some of these results are supported by observations made on bobwhite quail chicks, which, being younger, are thought to be more susceptible to the effects of ammonium perchlorate (AP) [79]. In a very carefully designed study using a wide range of dose levels in drinking water (12 treatment groups, ranging from 0.013 to 4,000 mg/L ammonium perchlorate), McNabb et al. hypothesized (1) that AP exposure will decrease plasma T_4 concentrations, increase thyroid gland weight, and decrease thyroidal T_4 content in quail, and (2) that quail will show some degree of adaptation or compensation in their thyroid function in response to sustained AP exposure. They concluded that "With sustained AP exposure (8 weeks), at the lowest range of AP concentrations used, chicks showed adaptation in thyroid function that fully compensated for the initial (2 week) effects of AP. At the intermediate AP concentrations there was partial compensation for the initial AP effects. At the highest AP exposures used, thyroid function was very low throughout the study, with no indication of compensatory responses. The capability of chicks to increase some aspects of their thyroid function adaptively in response to some levels of sustained AP exposure is contrary to the common generalization that developing animals are most vulnerable to environmental contaminants." The apparent LOEL is 50 μg/L AP, while the NOEL appears to be 25 μg/L.

In a SERDP report [76] the same authors evaluated low dose exposure of mallard ducklings to AP via drinking water and found an LOEL of 5 mg/L and an NOEL of 1 mg/L (2 week exposure). Their data suggest that "thyroid function in ducklings is more resistant to AP effects than is the case for quail" (the latter having an LOEL of 50 μg/L). These bird studies support the supposition that the thyroid, being an endocrine gland, is able to readily adapt, via the hypothalamic/pituitary axis, to agents that may affect normal levels of iodide.

Carnivores

Predatory carnivores and or omnivores can vary in size (and activity) from the masked shrew to the red fox. Though perchlorate has been shown to bioconcentrate in many types of plants, such as grasses, shrubs, cactus, and trees, there have yet to be any studies that demonstrate that perchlorate will bioaccumulate or biomagnify up the food web (and most water soluble substances will not).

5.2.3   Risk Characterization

In a quantitative ecological risk assessment, the hazard that perchlorate poses is determined by dividing the dose the animal is exposed to in the environment (often called the Expected Environmental Concentration, or EEC) by the safe (or acceptable) dose (Toxicity Reference Value, or TRV, usually derived from animal bioassays). If the hazard quotient from this calculation is less than one, then no adverse effect is expected. If the hazard quotient is more than one, then the possibility of an adverse effect occurring increases, but because TRVs err on the safe (conservative) side, the hazard still may be low (or negligible).

Quantitative ecological risk assessments are generally site-specific and therefore outside the realm of this book. Table 5.10, however, attempts to put the hazard of perchlorate into perspective for individual media and types of wildlife, albeit in qualitative terms. Although this table is based on the careful review of available scientific studies available at the time of this writing, it is important to note that information on perchlorate, and how we perceive the risks, changes with time.

The literature on the effects of perchlorate exposure on ecological receptors indicates that intimate contact and frequent exposure to relatively high concentrations of perchlorate in soil (mg/kg), food (mg/kg) and/or drinking water (>50 μg/L) need to be regularly maintained in order to induce a change in thyroid function, as evidenced by a decrease in either stored or circulating thyroid hormone, or, alternatively, an increase in circulating TSH. These types and levels of exposures can only occur at sites that have moderate to very high concentrations of perchlorate in soil and/or surrounding habitat.

The class of animals at the highest risk from frequent exposure would be, according to U.S. EPA [8], herbivores and/or soil-dwelling mammals. Even so, the results from analyzing wildlife tissue samples at perchlorate-contaminated sites appear to be mixed (e.g., raccoons, mice, rats and shrews at LHAAP [58]; mice tested at the Massachusetts Military Reservation [60]). This may be due to the fact that (1) small mammal foraging activity is somewhat selective and/or (2) perchlorate, being water soluble and resistant to metabolic oxidation or reduction, is effectively "cleared" from the previously exposed animal, via the kidneys and urine. Additionally, in deciduous areas, decaying leaf litter, over a number of seasons, may accrue to such an extent as to "bury" soil horizons that were previously contaminated with perchlorate, thus reducing exposure potential.

5.3 Summary and Conclusions

This chapter discussed the state of risk assessment science in determining the potential hazards to human health and the environment from perchlorate contamination. Aquatic organisms are primarily exposed to perchlorate in surface water. Because microorganisms in anaerobic sediments can reduce perchlorate, contamination in sediment may not be an important issue for aquatic organisms.

For humans and other terrestrial organisms, exposure primarily occurs through drinking water but may also occur indirectly through consumption of vegetation, dairy products, or breast milk. Due to the low volatility of perchlorate, it does not form a vapor that can be inhaled. Because perchlorate does not accumulate in animal tissues, due to its solubility, carnivores are not exposed to perchlorate through meat consumption.

Once ingested, perchlorate is rapidly absorbed, has a short half-life in the human body (approximately 8 hours), and is rapidly excreted unchanged in the urine. The primary concern for humans and other higher-order mammals is the potential for perchlorate to competitively inhibit iodine uptake by the thyroid gland. This effect reverses completely when exposure to perchlorate ceases. No evidence suggests that low doses of perchlorate cause thyroid disorders, thyroid nodules or cancer of the thyroid gland, nor has it been shown to cause adverse effects in any other organ. It is also not genotoxic or mutagenic.

Table 5.10 Anticipated Hazard Potential for Selected Ecological Receptors at the Perchlorate-Contaminated Sites

| | POTENTIAL EXPOSURE/HAZARD OF ENVIRONMENTAL RECEPTORS THAT ARE: | | | | | | |
| | AQUATIC | | SEMIAQUATIC | | TERRESTRIAL | | |
	AQUATIC INVERTEBRATES	FISH	WADING BIRDS	REPTILES & AMPHIBIANS	TERRESTRIAL INVERTEBRATES	PERCHING BIRDS OR RAPTORS	MAMMALS
Soil		Incomplete Exposure Pathway	LOW HAZARD to birds based on testing of avian species (e.g., quail) in laboratory; bioaccumulation potential very low	LOW HAZARD: weak partitioning of perchlorate to sediment; soil exposure anticipated to be minimal	LOW HAZARD based on bioassays conducted on invertebrates exposed to soil containing perchlorate	LOW HAZARD to birds based on testing of avian species (e.g., quail) in laboratory; biomagnification potential very low	MODERATE HAZARD based on intimate contact; LOW HAZARD to large mammals or ungulates based on extensive home range and large mass
Surface water	LOW HAZARD based on freshwater aquatic bioassays (s ee Section 6.1.4 on derivation of Ambient Water Quality Criteria)	NEGLIGIBLE HAZARD: low bioconcentration in fish; surface water concentrations rarely elevated		MODERATE HAZARD: apparent sensitivity of some types of amphibians to perchlorate in lab	Incomplete exposure pathway	NEGLIGIBLE HAZARD : surface water concentrations rarely elevated (e.g. , >i g/L range) to pose risk as a drinking water source	

Table 5.10 Anticipated Hazard Potential for Selected Ecological Receptors at the Perchlorate-Contaminated Sites, continued

	POTENTIAL EXPOSURE/HAZARD OF ENVIRONMENTAL RECEPTORS THAT ARE:						
	AQUATIC		**SEMIAQUATIC**		**TERRESTRIAL INVERTEBRATES**	**TERRESTRIAL**	
	AQUATIC INVERTEBRATES	**FISH**	**WADING BIRDS**	**REPTILES & AMPHIBIANS**		**PERCHING BIRDS OR RAPTORS**	**MAMMALS**
Sediment	LOW HAZARD based on the fact that perchlorate will primarily partition to surface water; surface water concentrations rarely elevated (e.g. , > i g/L range)				Incomplete exposure pathway	Incomplete exposure pathway	
Ground water	NEGLIGIBLE HAZAR D: groundwater rarely in mg/L range; subsurface flow to surface water also considered to be negligible				Incomplete exposure pathway	NEGLIGIBLE HAZARD: contribution of groundwater flow to surface water considered to be negligible	
Air	NEGLIGIBLE HAZARD: D eposition perchlorate contaminated dust anticipated to be low.		NEGLIGIBLE HAZARD: inhalation of perchlorate contaminated dust anticipated to be low.		Incomplete exposure pathway	NEGLIGIBLE HAZARD: inhalation of perchlorate-contaminated dust anticipated to be low.	

In a summary of toxicological effects, the NAS [5] states "An important point is that inhibition of thyroid iodide uptake is the only effect that has been consistently documented in humans exposed to perchlorate." The NAS [5] also cautioned, however, that: "no studies have investigated the relationship between perchlorate exposure and adverse outcomes among especially vulnerable groups, such as low-birth weight or preterm infants." The NAS further noted that the epidemiologic evidence to date indicates no causal association between prenatal exposure of perchlorate in drinking water (to concentrations as high as 100 to 120 μg perchlorate per liter of drinking water) during gestation and changes in newborn thyroid hormone levels or congenital hypothyroidism rates. Furthermore, no study to date has reported any abnormality in growth or thyroid function in children similarly exposed to perchlorate in drinking water.

Some animal studies have shown some endocrine disruption effects, such as the inhibition of forelimb emergence and tail resorption in frogs. These effects are completely reversible when exposure to perchlorate ceases. It is not clear at this time how such effects may affect wild populations of aquatic organisms. Endocrine disruption has not been observed in humans exposed to perchlorate.

Regulators use the results of research into toxicological effects and evaluations of potential exposures as the basis for deriving standards and guidance levels for perchlorate, which is described in Chapter 6.

5.4 References

[1] National Institutes of Health, National Cancer Institute, *Changes in Cigarette-Related Disease Risks and Their Implication for Prevention and Control, Monograph 8*, Washington D.C., National Institutes of Health, 1997.

[2] Thompson, R.S., Rivara, F.P., and Thompson, D.C., "A case-control study of the effectiveness of bicycle safety helmets," *New England Journal of Medicine*, 320, 1361-1367, 1989.

[3] National Research Council (NRC) of the National Academy of Science (NAS), *Risk Assessment in the Federal Government, Managing the Process,* National Academy Press, Washington, D.C., 1983.

[4] U.S. EPA, Risk Assessment Guidance for Superfund, Volume I: Human Health Evaluation Manual (Part A), EPA/540/1-89-002, December 1989.

[5] National Research Council (NRC) of the National Academy of Science (NAS), *Health Implications of Perchlorate Ingestion*, Committee to Assess the Health Implications of Perchlorate Ingestion, National Academies Press, January 2005.

[6] American Water Works Association, Perchlorate Occurrence Mapping, report prepared by Brandhuber, P. and Clark, S., HDR, Government Affairs Office, 1401 New York Ave., NW, Suite 640, Washington, D.C. 20005, January 2005.

[7] Memorandum from Susan Parker Bodine, Assistant Administrator, U.S. EPA Office of Solid Waste and Emergency Response to Regional Administrators, Subject: Assessment Guidance for Perchlorate, January 26, 2006.

[8] U.S. EPA., *Perchlorate Environmental Contamination: Toxicological Review and Risk Characterization*, Office of Research and Development, External Review Draft NCEA-1-0503, January 16, 2002.

[9] ATSDR, Draft Toxicological Profile for Perchlorates, U.S. Department of Health and Human Services, Public Health Service, Agency for Toxic Substances and Disease Registry, Division of Toxicology and Environmental Medicine/ Toxicology Information Branch, 1600 Clifton Road NE, Mailstop F-32, Atlanta, GA, http://www.atsdr.cdc.gov/toxpro2.html#Draft, 2005.

[10] *U.S. EPA*, s.v. "Integrated Risk Information System [IRIS]," http://www.epa. gov/iris/subst/1007.htm#doccar (accessed February 18, 2005).

[11] *Norman Endocrine Surgery Clinic,* s.v. "How Your Thyroid Works: A Delicate Feedback Mechanism," www.endocrineweb.com/thyfunction.html.

[12] Crump, C. et al., "Does Perchlorate in Drinking Water Affect Thyroid Function in Newborns or School-Age Children?" *J. Occup. Environ. Med.*, 42(6), 603-612, 2000.

[13] Tellez, R.T. et al., Long-term environmental exposure to perchlorate through drinking water and thyroid function during pregnancy and the neonatal period, *Thyroid*, 15(9), 963, 2005.

[14] Buffler, P.A. et al., Thyroid function and perchlorate in drinking water: an evaluation among California newborns, *Environ. Health Perspect.*, 1998, doi:10.1289/ehp.8176, 2005, http://dx.doi.org/ (accessed December 15, 2005).

[15] Rockette, H.E. and Arena, V.C., Mortality Pattern of Workers in the Niagara Plant, Department of Biostatistics, Graduate School of Public Health, University of Pittsburgh, Pittsburgh, PA, June 1983.

[16] Gibbs, J.P. et al., Evaluation of a population with occupational exposure to airborne ammonium perchlorate for possible acute or chronic effects on thyroid function, *J. Occup. Environ. Med.*, 40(12), 1072-1082, 1998.

[17] Lamm, S.H. et al., Thyroid health status of ammonium perchlorate workers: a cross-sectional occupational health study, *J. Occup. Environ. Med.*, 41, 248, 1999.

[18] Braverman, L.E. et al., The effect of low dose perchlorate on thyroid function in normal volunteers, *Thyroid,* 14(9), 691, 2004.

[19] Brabant, G. et al., Early adaptation of thyrotropin and thyroglobulin secretion to experimentally decreased iodine supply in man, *Metabolism*, 41, 1093-1096, 1992.

[20] Lamm, S.H and Doemland, M., Has perchlorate in drinking water increased the rate of congenital hypothyroidism?, *J. Occup. Environ. Med.*, 41(5), 409, 1999.

[21] Li, Z. et al., Neonatal thyroxine level and perchlorate in drinking water, *J. Occup. Environ. Med.*, 42, 200, 2000.

[22] Li, F.X. et al., Neonatal thyroid-stimulating hormone level and perchlorate in drinking water, *Teratology*, 62, 429, 2000.

[23] Greer, M.A. et al., Health effects assessment for environmental perchlorate contamination: the dose response for inhibition of thyroidal radioiodine uptake in humans, *Environ. Health Perspect*, 110, 927, 2002.

[24] Lawrence, J.E. et al., The effect of short-term low-dose perchlorate on various aspects of thyroid function, *Thyroid*, 10(8), 659-663, 2001.

[25] Schwartz, J., Gestational Exposure to Perchlorate is Associated with Measures of Decreased Thyroid Function in a Population of California Neonates, M.S. Thesis, University of California, Berkeley, 2001.

[26] Brechner, R.J. et al., Ammonium perchlorate contamination of Colorado River drinking water is associated with abnormal thyroid function in newborns in Arizona, *J. Occup. Environ. Med.*, 42, 777-782, 2000.

[27] Chang, S. et al., Pediatric neurobehavioral diseases in Nevada counties with respect to perchlorate in drinking water: An ecological inquiry, *Birth Defects Res. Part A Clin. Mol. Teratol.*, 67(10), 886, 2003.

[28] Lamm, S.H., Perchlorate exposure does not explain differences in neonatal thyroid function between Yuma and Flagstaff, [Letter]. *J. Occup. Environ. Med.*, 45(11), 1131, 2003.

[29] Kelsh, M.A. et al., Primary congenital hypothyroidism, newborn thyroid function, and environmental perchlorate exposure among residents of a Southern California community, *J. Occup. Environ. Med.*, 45(10), 1116, 2003.

[30] *U.S. EPA*, s.v. "Data from the UCMR 1, covering the period 2001-2005," http://www.epa.gov/safewater/ucmr/data.html#ucmr1 (accessed January 26, 2006).

[31] *U.S. EPA*, s.v. "Known Perchlorate Releases in the U.S. - March 25, 2005," http://clu-in.org/contaminantfocus/default.focus/sec/perchlorate/cat/Environmental Occurrence/#n1 (accessed January 2006).

[32] *Massachusetts Department of Environmental Protection,* s.v. "The Occurrence and Sources of Perchlorate in Massachusetts Draft Report," http://www.mass.gov/dep/cleanup/sites/percsour.pdf (accessed August 2005).

[33] Jackson, W.A. et al., Distribution and Potential Sources of Perchlorate in the High Plains Region of Texas, Final Report, Texas Tech University Water Resources Center, August 31, 2004.

[34] Hogue, C., Rocket-Fueled River, *Chemical and Engineering News,* 81, 33, 2003, http://pubs.acs.org/cen/coverstory/8133/8133perchlorates.html.

[35] U.S. EPA, *Perchlorate Monitoring Results - Henderson, Nevada to the Lower Colorado River, June 2005 Report,* Compiled by USEPA, Region 9, Waste Management Division, July 25, 2005.

[36] United States Geological Survey (USGS), National Stream Quality Accounting Network, 2000, http://water.usgs.gov/nasqan/progdocs/factsheets/clrd fact/clrdfact.html.

[37] United States Food and Drug Administration (FDA), *Exploratory Data on Perchlorate in Food,* Office of Plant and Dairy Foods, November 2004.

[38] Ellington, J.J. et al., Determination of Perchlorate in Tobacco Plants and Tobacco Products, *Environ. Sci. Technol.,* 35, 3213-3218, 2001.

[39] Yu, L. et al., Uptake of perchlorate in terrestrial plants, *Ecotoxicology and Environmental Safety,* 58, 44-49, 2004.

[40] Jackson, W.A. et al., Perchlorate accumulation in forage and edible vegetation, *J. Agric. Food Chem.,* 53, 369-373, 2005.

[41] Sanchez, C.A. et al., Perchlorate and Nitrate in Leafy Vegetables of North America, *Environ. Sci. Technol.,* 39(24), 9391-9397, 2005.

[42] Tan, K. et al., Accumulation of Perchlorate in Aquatic and Terrestrial Plants at a Field Scale, *J. Environ. Qual.,* 33, 1638-1646, 2004.

[43] Cheng, Q. et al., A Study on Perchlorate Exposure and Absorption in Beef Cattle, *J. Agric. Food Chem,* 52, 3456-3461, 2004.

[44] Utah Department of Agriculture and Food (UDAF), Perchlorate Fact Sheet, December 14, 2004, http://www.ag.state.ut.us/pressrel/perchloratefacts2.html.

[45] Kirk, A.B. et al., Perchlorate in Milk, *Environ. Sci. Technol.,* 37, 4979-4981, 2003.

[46] Kirk, A.B. et al., Perchlorate and Iodide in Dairy and Breast Milk, *Environ. Sci. Technol.,* 39, 2011-2017, 2005.

[47] U.S. Food and Drug Administration (FDA), Perchlorate Questions and Answers, U.S. Food and Drug Administration, Center for Food Safety and Applied Nutrition, CFSAN/Office of Plant & Dairy Foods, September 20, 2003, Updated November 26, 2004, http://www.cfsan.fda.gov/~dms/clo4qa.html (accessed November 11, 2005).

[48] The Interstate Technology and Regulatory Council (ITRC) Perchlorate Team, Perchlorate: Overview of Issues, Status, and Remedial Options, September 2005.

[49] Greer, M.A. et al., Health effects assessment for environmental perchlorate contamination: the dose response for inhibition of thyroidal radioiodine uptake in humans, *Environ. Health Perspect.,* 110, 927-937, 2002.

[50] Parsons Engineering Science, Inc., Scientific and Technical Report for perchlorate biotransport investigation: a study of perchlorate occurrence in selected ecosystems, Interim final, Austin, TX, Contract no. F41624-95-D-9018, 2001.

[51] Smith, P.N. et al., Perchlorate in water, soil, vegetation, and rodents collected from the Las Vegas Wash, Nevada, *Environmental Pollution,* 132, 121-127, 2004.

[52] Theodorakis, C. et al., Perchlorate in fish from a contaminated site in east-central Texas, *Environmental Pollution,* 139, 59-69, 2006.

[53] Tan, K., Anderson, T.A., and Jackson, W.A., Degradation kinetics of perchlorate in sediments and soils, *Water, Air, and Soil Pollution,* 151, 245-259, 2004.

[54] Tan, K., Anderson, T.A., and Jackson, W.A., Temporal and spatial variation of perchlorate in streambed sediments: results from in-situ dialysis samplers, *Environmental Pollution,* 136, 283-291, 2005.

[55] AMEC Earth and Environmental, Inc., Personal communication, Nelson, A. to Clough, S., Access database query of 01/05/06 for descriptive statistics of soil samples analyzed for perchlorate, Massachusetts Military Reservation, Bourne, MA, January 2006.

[56] Orris, G.J. et al., Preliminary analyses for perchlorate in selected natural materials and their derivative products, U.S. Department of Interior, U.S. Geological Survey, Open-File Report 03-314, 2003, http://geopubs.wr.usgs.gov/open-file/of03-314/OF03-314.pdf.

[57] Harvey, G. et al., Plant perchlorate accumulation, [Poster], Partners in Environmental Technology, Technical Symposium Workshop, Washington, D.C. Nov. 29-Dec 1, 2005.

[58] Smith, P.N. et al., Preliminary assessment of perchlorate in ecological receptors at the Longhorn Army Ammunition Plant (LHAAP), Karnack, Texas, *Ecotoxicology*, 10, 305-313, 2001.

[59] Harvey, G. et al., Plant perchlorate accumulation, Poster #39, Partner in Environmental Technology Technical Symposium and Workshop, Washington, D.C. November 29-December 1, 2005.

[60] Alsop, W.R., Samuelian, J.H., and Davis, R., Assessing Alternate Approaches to Estimating Uptake of Compounds by Plants and Animals in Ecological Risk Assessments, The 19th Annual International Conference on Soils, Sediments, and Water, University of Massachusetts, Amherst, October 20-23, 2003.

[61] U.S. EPA, Final Water Quality Guidance for the Great Lakes System: Final Rule 60FR15365, 1995.

[62] U.S. EPA, Great Lakes Water Quality Initiative Tier II water quality values for protection of aquatic life in ambient water, Support documents, November 23, 1992.

[63] U.S. EPA, Disposition of Comments and Recommendations for Revisions to "Perchlorate Environmental Contamination: Toxicological Review and Risk Characterization External Review Draft (January 16, 2002)," EPA/600/R-03/031, October 2003.

[64] Dean, K.E. et al., Development of freshwater water-quality criteria for perchlorate, *Environmental Toxicology and Chemistry*, 23(6), 1441-1451, 2004.

[65] Tietge, J.E. et al., Metamorphic inhibition of Xenopus laevis by sodium perchlorate: effects on development and thyroid histology, *Environmental Toxicology and Chemistry*, 24(4), 926-933, 2005.

[66] Goleman, W.L. et al., Environmentally relevant concentrations of ammonium perchlorate inhibit development and metamorphosis in Xenopus laevis, *Environmental Toxicology and Chemistry*, 21(2), 424-430, 2002.

[67] Goleman, W.L., Carr, J.A., and Anderson, T.A., Environmentally relevant concentrations of ammonium perchlorate inhibit thyroid function and alter sex ratios in developing Xenopus laevis, *Environmental Toxicology and Chemistry*, 21(3), 590-597, 2002.

[68] Carr, J.A. et al., Response of native adult and larval anurans in their natural environment to ammonium perchlorate contamination: Assessment of reproductive and thyroid endpoints, Final report submitted to the Strategic Environmental Research and Development Program (SERDP), Lubbock, TX, The Institute of Environmental and Human Health (TIEHH), Texas Tech University, March 28, 2002.

[69] Smith, P.N. et al., Monitoring perchlorate exposure and thyroid hormone status among raccoons inhabiting a perchlorate-contaminated site, *Environ. Monit. Assessment*, 102, 337-347, 2005.

[70] EA Engineering, Science, and Technology, Inc., Results of acute and chronic toxicity testing with sodium perchlorate, Brooks Air Force Base, TX, Armstrong Laboratory; report no. 2900, 1998.

[71] Anderson, T.A., Uptake of the perchlorate anion into various plant species, Final report CU 1223, Submitted to the Strategic Environmental Research and Development Program (SERDP), Lubbock, TX, Texas Tech University, The Institute of Environmental and Human Health (TIEHH), Project no. T9700, March 28, 2002.

[72] Argus Research Laboratories, Inc., *Hormone, Thyroid and Neurohistological Effects of Oral (Drinking Water) Exposure to Ammonium Perchlorate in Pregnant and Lactating Rats and in Fetuses and Nursing Pups Exposed to Ammonium Perchlorate During Gestation or Via Maternal Milk,* ARGUS 1416-003, Argus Research Laboratories, Inc., Horsham, PA, 2001.

[73] Thuett, K.A. et al., In utero and lactational exposure to ammonium perchlorate in drinking water: effects on developing deer mice at postnatal day 21, *J. Toxicol. Environ. Health,* 65, 1061-1076, 2002.

[74] Isanhart, J.P., McNabb, F.M.A., and Smith, P.N., Effects of perchlorate on resting metabolism, peak metabolism, and thyroid function in the prairie vole, *Environ. Toxicol. Chem.,* 24, 678, 2005.

[75] Gentles, A., Surles, J., and Smith., E.E., Evaluation of adult quail and egg production following exposure to perchlorate-treated water, *Environ. Toxicol. Chem.,* 24(8), 1930-1934, 2005.

[76] SERDP, *The effects of perchlorate on developing and adult birds,* SERDP Project CU-1242, F.M.A. McNabb, Investigator, October 2001-June 2003.

[77] Argus Research Laboratories, Inc., *A neurobehavioral developmental study of ammonium perchlorate administered orally in drinking water to rats,* ARGUS 1613-002, Argus Research Laboratories, Inc., Horsham, PA, 1998.

[78] Argus Research Laboratories, Inc., *Hormone, thyroid and neurohistological effects of oral (Drinking Water) exposure to ammonium perchlorate in pregnant and lactating rats and in fetuses and nursing pups exposed to ammonium perchlorate during gestation or via maternal milk,* ARGUS 1416-003, Argus Research Laboratories, Inc., Horsham, PA, 2001.

[79] McNabb, F.M.A., Jang, D.A., and Larsen, C.T., Does thyroid function in developing birds adapt to sustained ammonium perchlorate exposure?, *Tox. Sci.,* 82, 106-113, 2004.

[80] Histology graphic courtesy of Dr. Roger C. Wagner, University of Delaware: http://www.udel.edu/Biology/Wags/histopage/histopage.htm. With permission.

[81] Simon, R.T. and Weber, E.J., Reduction of perchlorate in river sediment, *Envir. Toxicol. Chem.,* 25(4), 899-903, 2006.

CHAPTER 6

What Are the Regulatory Limits on Perchlorate?

The preceding chapter described the toxicological data, the potential for human and environmental exposure to perchlorate, and how the U.S. EPA has developed screening levels for perchlorate in the environment based on those data. Screening levels do not, however, carry the weight of a promulgated regulation. This chapter describes developing regulatory requirements for perchlorate under the Safe Drinking Water Act and related risk management efforts.

Definitions
- Acute - Short term (e.g., Ambient Water Quality Criterion based on 96-hour toxicity testing)
- Ambient Water Quality Criteria (AWQC) - Numeric limits on the amounts of chemicals that can be present in surface water without harm to aquatic life such as death, slower growth, reduced reproduction, and the accumulation of levels of toxic chemicals in their tissues that may harm consumers of those organisms.
- Chronic - Long term; in the context of aquatic toxicity testing, typically means from seven days to full life cycle
- Maximum Contaminant Level (MCL) - The highest level of a contaminant allowed in a public drinking water supply, considering potential short-term or long-term health risks and the economical and technological feasibility of treatment. MCLs are legally enforceable standards.
- Maximum Contaminant Level Goal (MCLG) - The level of a contaminant in drinking water at which there would be no risk to human health. MCLGs are not legally enforceable.
- Relative Source Contribution (RSC) - Factor used in developing a MCL to account for intake of contaminants from sources other than drinking water.

6.1 Risk Management and Regulatory Limits

Risk management decisions include the determination of regulatory limits and non-enforceable guidance levels. Decision makers use the information from a risk assessment in the context of other scientific and non-scientific factors to develop strategies to minimize or eliminate risks due to chemical exposures. The risk managers and their decisions may differ by region, state, and by the particular issue at hand. Often, risk management decisions vary significantly depending upon regulatory requirements and differences in interpreting the risk assessment, as well as other scientific, social, economic, and other non-scientific issues associated with regulating public health limits, placing risk managers in unenviable positions.

At the federal level, the Safe Drinking Water Act authorizes the U.S. EPA as the decision maker to set enforceable drinking water standards, also referred to as Maximum Contaminant Level (MCLs). Individual state environmental agencies may choose to develop their own more stringent drinking water standard or adopt the federal MCL. In general, drinking water standards are derived for contaminants that are found to exist in public water supplies at concentrations that pose a likely public health concern. Furthermore, an effective treatment method to reduce the contaminant's concentration must exist in order for the U.S. EPA or a state to set a drinking water standard. For perchlorate, several scientific issues have affected the development of drinking water standards, including the availability of analytical methods to truly determine if perchlorate is present, the determination of what constitutes a level of significant risk, the determination of how prevalent perchlorate exists in both drinking water and in other environmental media, and the existence of suitable treatment options.

Under the federal Comprehensive Environmental Response, Compensation & Liability Act (CERCLA) of 1980 as amended in 1986 (also referred to as Superfund) and the U.S. Resource Conservation and Recovery Act (RCRA) of 1976, risk managers include regional U.S. EPA and/or state regulators, as well as residents or other individuals who have an interest in the decisions (stakeholders). These risk managers determine soil and groundwater clean-up goals for hazardous waste sites.

Under the Clean Water Act, the U.S. EPA is required to develop and publish water quality criteria to protect designated uses of surface water bodies. Similar to drinking water standards, states may choose to adopt these criteria as enforceable standards or develop their own. Unlike drinking water standards, however, the federal water quality criteria are based solely on data and scientific judgments on the relationships between pollutant concentrations and environmental and human health effects; economic impacts or technological feasibility of meeting the criteria concentrations are not considered when developing the water quality criteria. The risk management process in developing the drinking water, clean-up goals and water quality criteria for perchlorate is presented in the remainder of this chapter.

6.1.1   Safe Drinking Water Act

The U.S. EPA regulates compounds in drinking water under the Safe Drinking Water Act (SDWA). No national drinking water regulations currently exist for perchlorate. In 1998, the U.S. EPA added perchlorate to the Drinking Water Contaminant Candidate List (CCL) based on its presence in drinking water supplies in the southwestern United States [1]. On February 24, 2005, the U.S. EPA included perchlorate on the second CCL. This designation provides for comprehensive studies regarding analytical methods for detecting the contaminant, the prevalence of occurrence in drinking water, potential health effects, and the efficacy of treatment technologies to remove the contaminant from drinking water. Based on the information from these studies, the U.S. EPA will formally determine whether it should issue a national primary drinking water regulation for perchlorate. In accordance with Federal environmental laws, the U.S. EPA will make a final regulatory determination regarding the second CCL in August of 2006 [2].

Under the SDWA, the U.S. EPA must determine that the regulation of perchlorate represents a meaningful mechanism or opportunity to reduce health risks. As part of the determination of whether perchlorate is a public health issue, the Agency must simultaneously consider the following:

- How widespread is the occurrence of perchlorate in public water systems?
- Does perchlorate exposure result in adverse health effects?
- What is the availability and associated cost of treatment technology?
- Do technologically defensible analytical methods exist for a range of different water matrices to measure perchlorate at whatever the ultimate drinking water standard may be?
- Would a proposed drinking water standard present a meaningful opportunity to reduce health risks associated with perchlorate?

To answer the question of how widespread is the occurrence of perchlorate in public water systems, the U.S. EPA is monitoring perchlorate using analytical Method 314.0 (see Chapter 3) under the Unregulated Contaminant Monitoring Rule (UCMR) (40 CFR Parts 9, 141, 142). This federal program monitors perchlorate and other CCLs in surface water and groundwater sources for large and small public water systems. By definition, large public water systems serve more than 10,000 people while small systems serve 10,000 people or less [2].

Under the first phase of the UCMR, which was conducted between 2001 and 2005, perchlorate was monitored in water samples from approximately 2,800 large public water systems and 800 out of a total of 66,000 small systems, representing a statistical sample of small systems. As of January 2005, the data from the first UCMR survey found perchlorate detected in 153 public water systems in 25 states, with detections most prevalent in southern California, west central Texas, the east coast, represented by Massachusetts, and the area between New Jersey and Long Island, New York (see Table 5.3). In addition to the U.S. EPA's efforts, various state agencies, the United States Geological Survey, and several large university-led research projects are currently collecting data in support of public health determinations of the prevalence of perchlorate in water supplies as well as other sources (e.g., food).

In general, the U.S. EPA attributes the presence of perchlorate in water to facilities that manufactured, tested, or disposed of solid rocket fuel, fire works, flares and other industrial sources of perchlorate. However, the source of perchlorate in all areas is not known and likely includes natural sources. The U.S. EPA has proposed a second round of water sampling beginning in 2007 and continuing through 2011 for perchlorate and other CCLs under the UCMR [2].

As described in Chapter 5, regulators and other researchers have studied the potential for adverse health effects from perchlorate exposure for many years. In February 2005, the U.S. EPA established a final perchlorate reference dose of

0.0007 milligrams perchlorate per kilogram body weight per day (expressed as mg/kg•day) [3]. This reference dose, which was also recommended by the National Research Council (NRC) of the National Academy of Sciences (NAS) [4], represents the consensus position among health professionals from a variety of perspectives on the health effects associated with chronic exposure to perchlorate. The reference dose represents an oral dose that a person, including sensitive subgroups such as an expectant mother, fetus, and newborn, could be exposed to on a daily basis over a lifetime without adverse health effects [3].

6.1.1.1 MCLs and MCLGs

The reference dose for perchlorate serves as the health-effects basis of the rule-making process under the SDWA to establish MCLs. This process begins with the calculation of a Maximum Contaminant Level Goal, which is based on the Drinking Water Equivalent Level (DWEL). After considering technical limitations on monitoring and water treatment, the Agency may then promulgate an MCL. The U.S. EPA is currently deciding whether to promulgate a Federal MCL for perchlorate. Should that rulemaking proceed, it will follow the process described in more detail below. In general, the U.S. EPA is required by law to issue a proposed regulation two years after determining that the regulation is necessary, with a final regulation promulgated two years later. This process can be expedited if a contaminant presents an urgent threat to public health.

After reviewing available health effects studies, the U.S. EPA sets a Maximum Contaminant Level Goal (MCLG), which is the maximum level of a contaminant in drinking water at which no known or anticipated adverse effect on human health would occur, allowing for an adequate margin of safety. MCLGs are non-enforceable public health goals. When determining an MCLG, the U.S. EPA considers the risk to sensitive subpopulations, such as infants, children, the elderly, and those with compromised immune systems, of experiencing a variety of adverse health effects. Since the derivation of MCLGs considers only public health issues and not the limits of detection and treatment technologies, often times MCLGs are set at levels that are not technologically feasible for water systems to attain.

For compounds such as perchlorate that are not considered a human carcinogen, the MCLG is based on the reference dose multiplied by a typical male adult body weight (70 kg) and divided by daily water consumption for a sedentary adult (2 liters) to provide a DWEL. Based on the final reference dose, the U.S. EPA has calculated a DWEL of 24.5 μg/L for perchlorate [5]. Assuming that a population's entire dose of perchlorate comes from drinking water, the DWEL represents the concentration of perchlorate in drinking water that will have no adverse effect. Because the reference dose and the DWEL include added margins of safety in their derivation, exposures to perchlorate above 24.5 μg/L are not necessarily unsafe. State and federal regulators may use the DWEL to make decisions regarding groundwater clean-up at hazardous waste sites or, absent a promulgated MCL, decisions regarding perchlorate in drinking water.

Reference Dose (RfD) and Drinking Water Equivalent Level (DWEL)

The U.S. EPA develops a RfD [mg/kg•day] to represent the safe oral dose of a noncarcinogenic chemical for any human receptor, conservatively based on toxicological data and incorporating factors of safety (see Chapter 5).

Risk managers use this safe dose to calculate a Drinking Water Equivalent Level [DWEL], the concentration of a contaminant in drinking water that will have no adverse effect. The DWEL derived by U.S. EPA [5] assumes that a 70 kg adult drinks 2 L water per day with no exposure from other sources. The DWEL for perchlorate in drinking water is derived as follows:

$$DWEL = (0.0007 \text{ mg/kg} \cdot \text{day}) \times (70 \text{ kg}) \times (1 \text{ day/2 L}) = 0.0245 \text{ mg/L}$$

No other uncertainty factors need to be applied because laboratory animals were not used to obtain the NOEL, and no adjustments for acute-to-chronic dosing needed to be made.

Once the MCLG is determined, the U.S. EPA sets an enforceable standard which, in most cases, is an MCL. The MCL represents the maximum permissible level of a contaminant in water delivered to any user of a public water system. The MCL is set as close to the MCLG as feasible. The SDWA defines the MCL as the level that may be achieved using the best available technology, treatment techniques, and other means. The U.S. EPA evaluates these options after examining their potential efficiency under field conditions and not solely under laboratory conditions. The evaluation also considers the economic cost associated with each of these technologies into consideration when setting the MCL.

Although the U.S. EPA may set an MCL based upon the reference dose, individual states retain the right to set state standards if they so choose, including standards that may be stricter than federal standards. Since the U.S. EPA established the reference dose for perchlorate, three state agencies have proposed drinking water standards for perchlorate. Each of these states anticipate promulgating a drinking water standard for perchlorate in 2006.

California has reaffirmed a Public Health Goal[1] of 6 μg/L for perchlorate in drinking water based on a state-derived reference dose (0.0003 mg/kg•day) and an adult drinking water rate of 2 liters of water per day. It also includes the assumption that 60% of daily exposure to perchlorate comes from drinking water, accounting for other sources of perchlorate exposures [7].

New Jersey has also proposed an MCL for perchlorate of 5 μg/L. This value is based on U.S. EPA's reference dose (0.0007 mg/kg•day), an adult drinking 2 liters of water per day, and an assumption that tap water accounts for only 20% of an adult's daily exposure to perchlorate [8].

Finally, Massachusetts has proposed an MCL for perchlorate of 2 μg/L based on an interim drinking water guidance level and other considerations. Massachusetts

---

[1] California's Public Health Goal (PHG) is comparable to an MCLG. Two liters of drinking water per day is the U.S. EPA recommended default drinking water ingestion rate for adults and represents the 84th percentile value among adults (i.e., 84% of the population drinks this amount of water or less per day [6]).

initially established an interim drinking water guidance level of 1 μg/L based on a state-derived reference dose (0.00007 mg/kg•day) [9], interpretative calculations to protect sensitive populations, including infants, who can be exposed to perchlorate through breast milk, and an assumption that tap water accounts for only 20% of a daily exposure to perchlorate [10]. Since the interim level was established, the state has been monitoring perchlorate levels in public water supplies and assessing potential sources. Massachusetts has determined that hypochlorite solutions used in the disinfection processes at water treatment plants could result in detectable levels of perchlorate (approximately 1 μg/L) in chlorinated drinking water distribution systems [11]. Regulators considered this information in light of the potential economic and public health impacts on public water systems to comply with a 1 μg/L MCL for perchlorate. As a result, the risk managers decided an MCL of 2 μg/L was "extraordinarily protective" while ensuring other public health considerations remain intact [13].

6.1.1.2   Health Advisories

When no drinking water standard has been promulgated, a health advisory (HA) represents a non-enforceable guideline set by federal or state regulators to evaluate the health significance of a contaminant in drinking water. A health advisory represents the concentration of a particular contaminant that can be consumed daily in drinking water over different durations of exposure (i.e., one-day, ten-day, and lifetime) assuming drinking water is the only source of exposure. If drinking water concentrations exceed health advisory levels, then regulators require certain actions, including increased monitoring, notification of consumers and local public health agencies and possibly removing that water source from the public water supply.

Nine states have issued health advisories for perchlorate as of December 2005, ranging from 1 to 18 μg/L [12]. The range of values reflects differences in interpretation of the toxicological information, the acknowledgment by some of the relative contribution from sources other than groundwater for human exposure to perchlorate (such as in food), and interpretations of what human subpopulation should be protected (e.g., unborn fetus, persons with iodine deficiencies or hypothyroid disease). Table 6.1 presents a summary of the existing state health advisory levels for perchlorate in groundwater.

6.1.1.3   Variations in Drinking Water Standards and Advisories

Different approaches to deriving limits for perchlorate in drinking water can result in widely different requirements for treating drinking water around the country. These requirements can impose different costs for water treatment on utilities and consumers, or for groundwater remediation, in various states. Because of this consequence it is worth examining the regional differences in MCLs and health advisories in more detail.

Table 6.1 Summary of Proposed State Drinking Water Standards or Action Levels
for Perchlorate as of March 2006 [12] [13]

State/Other	PHG/MCLG (µg/L)	Proposed Drinking Water Standard (µg/L)	Action Level (µg/L)	Comments
Arizona			14	Based on child's exposures
Oregon			4	
California	6		6	Notification Level
New Jersey	5			
Maryland			1	
Massachusetts		2	1	
Nevada			18	
New Mexico			1	
New York			5	Drinking Water Planning Level
			18	Public Notification Level
Texas			17	Residential Protective Clean-up Level
			51	Industrial/Commercial Protective Clean-up Level
Canada			6	

Regulation of perchlorate in Massachusetts illustrates some important differences in regulating perchlorate and the consequences of those differences. In May 2004, the Massachusetts Department of Environmental Protection issued a final reference dose for perchlorate of 0.00007 mg/kg•day [9]. This preceded the U.S. EPA's adoption of the final reference dose of 0.0007 mg/kg•day [3]. The two reference doses reflect different interpretations of 1) the animal versus human toxicity studies, and 2) the application of uncertainty or safety factors deemed to be protective of sensitive subpopulations. Massachusetts derived an interim action level of 1 µg/L based on their reference dose, stating that this value protects sensitive subpopulations defined as pregnant women, infants, children up to age 12 and individuals with hypothyroidism [10]. Similarly, U.S. EPA has derived a DWEL of 24.5 µg/L based on their reference dose, stating the DWEL is appropriate and protective for all populations, including the most sensitive subgroups including the fetuses of pregnant women who might have hypothyroidism or iodide deficiency, premature neonates, infants and developing children [5].

Aside from different interpretations of the toxicity data, differences in assumptions regarding who is being exposed (adult or child) and associated drinking water ingestion rates and body weights, and differences in the relative source contribution (RSC) of drinking water exposures to all other potential sources of perchlorate exposures could yield a wide range of MCLs. For example, using U.S. EPA's reference dose, if one based the drinking water standard on an adult weighing 154 pounds (or 70 kilograms) who drinks 2 liters of water per day, the potential perchlorate drinking water standards range from 5 μg/L assuming a 20% relative source contribution, to 24.5 μg/L assuming 100% of perchlorate exposures come from drinking water. Although the reference dose is protective of the general population and sensitive subpopulations, an MCL could be derived based on alternative assumptions than those for an adult.[2]

The RSC term accounts for the fact that a population may be exposed to a chemical from sources other than their drinking water. Incorporating a RSC term into the derivation of a drinking water limit ensures that the total intake of a chemical will not result in exposures that exceed the reference dose [15]. To provide some consistency and guidance for considering aggregate exposures from all potential sources and the determination of an appropriate RSC, the U.S. EPA has developed an Exposure Decision Tree in an attempt to coordinate this policy across Agency programs [as cited in 15]. Under this framework, if data exist to estimate contributions from other sources, then those data can be used to allocate the RSC from drinking water. If no exposures other than drinking water are known to exist for a particular compound, then a conservative drinking water RSC of 80% is considered appropriate. If known but unquantifiable exposures other than drinking water exist, then a RSC value between 20 and 50% of the total daily exposure is attributed to drinking water. In practice, however, a value of 20% is typically used by state and federal agencies as a default allocation to exposures from drinking water [16]. Among the existing drinking water standards, the use of a 20% RSC value has been commonly applied as a default value. In other cases, the relative source contribution value varies for a single compound [15].

While a precise value for a perchlorate RSC cannot be established now, current scientific evidence suggests that the estimated exposure to perchlorate in water is greater than from other sources when considering the general adult population. Among the proposed drinking water standards and health advisories that currently exist for perchlorate, California used a RSC for groundwater of 60% [7], and New Jersey and Massachusetts used a RSC of 20% [8] [13].

Example of Relative Source Contribution of Perchlorate

Using standard U.S. EPA exposure assumptions and available information on perchlorate concentrations in potable water (assuming an average concentration of 12 μg/L [17]), and preliminary information regarding concentrations in food crops (assuming a maximum average by state of 10.4 μg/kg in lettuce [18] and an average concentration of 5.76 μg/L in milk, based on FDA data [18] [19] [20]), an adult would receive approximately 75% of total perchlorate exposure from drinking water, and 25% from food sources. Conversely, for a newborn fed exclusively formula reconstituted with tap water, the relative source contribution from water would be 100%.

---

[2] The federal MCL for nitrate is based on a 4 kg bottle-fed infant drinking 0.64 L of water per day [14].

General Population and Pregnant or Iodine Deficient Subpopulation	Ingestion via Drinking Water	Ingestion via Vegetables (lettuce)	Ingestion via Cow's Milk	Total Exposure (mg/Kg/day)
General Formula for Exposure: [Concentration] x Intake Rate/Body Weight x 0.001 µg/kg				
Perchlorate Concentration	12 µg/L	10.4 µg/kg	5.76 µg/L	
Intake Rate	2 L/day [8]	0.0285 (kg/day) [8]	0.71 L/day (3 x 8 ounce glasses per day)	
Body Weight	70 Kg [8]	70 Kg [8]	70 Kg [8]	
Pathway Specific and Total Exposure (mg/Kg/day)	3.43E-04	4.23E-06	5.84E-05	4.06E-04
Relative Source Contribution (Percent of Pathway-specific Exposure to Total Exposure for Adult	73.7%	1.8%	24.5%	

Note: Relative Source Contribution = Pathway-specific Exposure/Total Exposure

6.1.2 Clean-Up Goals under CERCLA and Analogous State Laws

In addition to drinking water standards, federal and state regulatory bodies determine clean-up goals for contaminants in soil, groundwater, and surface water. Where promulgated drinking water standards exist, such standards typically represent applicable or relevant and appropriate requirements and are often adopted as enforceable clean-up goals for groundwater. As no promulgated standards currently exist for perchlorate, different clean-up goals have been established for sites where perchlorate in groundwater must be remediated.

As noted in Table 6.1, Texas has issued a clean-up goal for perchlorate in groundwater of 51 µg/L for an industrial setting and 17 µg/L for residential settings. Massachusetts has also proposed generic site standards (termed "Method 1 Standards") derived in a manner to be protective of perchlorate exposures at a wide range of disposal sites across the state. The lowest Method 1 Standard for perchlorate in groundwater, 2 µg/L, is protective of drinking water exposures. The highest standard, 1,000 µg/L, was derived to be protective of potential environmental effects resulting from contaminated groundwater discharging to surface water [21]. In soil, the Method 1 Standards for perchlorate range from 0.1 mg/kg to 5 mg/kg based on the potential for exposure to soil [21].

Information regarding the range of site-specific groundwater clean-up goals was obtained from the U.S. EPA's list of federal facilities and Superfund sites where actions have been take to address perchlorate [16] and includes the following examples:

- Aerojet General Corp. Site - Rancho Cordova, CA (National Priorities List [NPL]), located fifteen miles east of Sacramento, California. In the July 2001 Record of Decision (ROD) for the first of multiple operable units (OUs) for the Aerojet site, the U.S. EPA set an enforceable site-specific clean-up goal for perchlorate at 4 μg/L in groundwater. This value also represented the Action Level in California.
- Hills Iowa Perchlorate Site - Hills, IA. In May of 2003, investigators confirmed detections of perchlorate above 4 μg/L in 21 residential drinking wells and one commercial business drinking water well that supplies approximately 115 employees. As an interim action, bottled water was supplied to those persons whose drinking water exceeded 18 μg/L of perchlorate, and to certain sub-populations, including pregnant women and women of child-bearing age, whose drinking water continued between 4 and 18 μg/L of perchlorate.
- Aberdeen Proving Grounds (APG), MD (NPL) - Approximately 36,000 people live within three miles of the site. Samples of groundwater from monitoring wells contained perchlorate ranging from 4 to 10 μg/L. Production well samples contained 1.0-2.1 μg/L perchlorate, and finished drinking water contained from 0.2-1.0 μg/L perchlorate, with an average perchlorate concentration in drinking water of 0.6 μg/L. As a result of the perchlorate contamination, the Maryland Department of the Environment issued an action level in drinking water of 1 μg/L. Prior to the U.S. EPA's adoption of a final reference dose for perchlorate in 2005, U.S. EPA Region III and the State had requested remediation of soil perchlorate in the range of 10 to 15 mg/kg based on previous understanding of perchlorate toxicity. Based on the final reference dose adopted in 2005, U.S. EPA Region III has established risk-based criteria for perchlorate in soils ranging from 55 mg/kg for residential scenarios to 720 mg/kg for industrial scenarios [22].
- Massachusetts Military Reservation, Camp Edwards, Cape Cod Massachusetts. The site overlies the Upper Cape Cod sole source aquifer. In 1997, an administrative order issued under the SDWA halted live training activities pending completion of investigations for the training area. Beginning in March 2002, perchlorate was detected in public supply wells within the town of Bourne's Monument Beach well field to the west of MMR. This well field contains four pumping wells that previously provided 70% of the town's drinking water. The Massachusetts Department of Environmental Protection issued an advisory to the town of Bourne stating that women, infants, children, and other sensitive sub-populations should not drink water that contains perchlorate above 1 μg/L [10]. More recently, Massachusetts has proposed a drinking water standard of 2 μg/L [13]. Until a federal or state MCL has been promulgated, DOD policy has established a perchlorate action level of 24 μg/L [23].

6.1.3   Clean Water Act

Acute and chronic Ambient Water Quality Criteria (AWQC) are the U.S. EPA's benchmarks for evaluating aquatic toxicity developed under the Clean Water Act. While regulators use AWQC to develop wastewater discharge permit limits, for example, they are not promulgated regulatory limits in and of themselves. The U.S. EPA develops AWQC through a formal protocol which includes data from a series of toxicity tests from a minimum dataset of at least eight different aquatic organisms.

Studies have shown that exposure to perchlorate at high concentrations can cause mortality in fish [24]. It has also been noted that exposure to perchlorate may delay metamorphosis in frogs, although the effect appears to be reversible. These findings were discussed previously in Section 5.2.2.1 and the derivation of benchmarks for water quality from these studies are described below.

U.S. EPA Protocol for Calculating AWQC [24]

Development of an AWQC begins with assembling acute aquatic toxicity test results that meet specific objectives, ranking the available toxicity test results, and calculating the lower fifth percentile of the distribution of toxicity test results. This value is used to develop the acute AWQC, and it can also be used to develop chronic AWQC. The data must include acute toxicity test results from at least eight different species to represent the potential range of sensitivities observed in the field.

Typically, these data consist of 96-hour LC_{50} data (concentrations lethal to 50% of the test organisms after a 96-hour exposure period). When more than one freshwater LC_{50} is available for a species within a genus, the geometric mean of all the data are calculated to represent the genus mean acute value (GMAV). The GMAVs are used to calculate the final acute value (FAV).

The FAV is an estimate of the concentration of a chemical corresponding to a cumulative probability of 0.05 in the distribution of acute toxicity values for aquatic species based on the range of acute toxicity test results for that chemical. The final acute value is divided by two to calculate the criterion maximum concentration (CMC). Because the data used are typically either lethal concentrations to 50% of the study organisms (LC_{50}) or effects concentrations to 50% of the study organisms (EC_{50}), the guidance [25] used a factor of two: "to result in a concentration that will not severely adversely affect too many of the organisms."

The chronic AWQC can be derived using the same protocols if chronic toxicity test data for eight species are available. However, U.S. EPA realized chronic aquatic toxicity test data are harder to come by because chronic toxicity testing is expensive and time consuming, and so developed a protocol to use more readily available acute toxicity test data with limited chronic toxicity test data. The protocol modifies the acute benchmark (FAV) based on the relationship between acute and chronic toxicity, the acute to chronic ratio (ACR). When that ratio is equal to one, acute and chronic effects are assumed to occur at the same concentration. When that ratio is ten, for example, chronic toxicity is assumed to occur at concentrations 1/10 of those at which acute effects are observed. The protocol requires species-specific results from three species with paired acute and chronic toxicity tests run in the same laboratory. The final ACR is calculated as the geometric mean of the three species-specific ACRs.

More recently, U.S. EPA [26] specified that when a species-specific ACR is not available to meet one of the requirements, the ACR should be assumed to be 18. The final acute value is divided by the final ACR to calculate the criterion continuous concentration (CCC).

In 2002, when U.S. EPA published its draft toxicological review and risk characterization for perchlorate [24], the aquatic effects of perchlorate had been evaluated in only a few studies where the concentration of the pure salt, typically sodium perchlorate or ammonium perchlorate, had been evaluated against the response observed following controlled application to a plant or animal. Because only two test results met the data requirements for calculating an AWQC, the U.S. EPA [24] concluded that the data would not suffice to calculate AWQC using the standard protocol. Instead the Agency calculated Tier 2 values using a protocol developed for the Great Lakes Water Quality Initiative [26] to calculate benchmark values for compounds with data that could not meet the strict criteria in the standard protocol. The resulting benchmark or Tier 2 values would be lower than the AWQC (were there sufficient data) in at least 80% of the cases [26] [27]. The Great Lakes Water Quality Initiative protocol generated acute and chronic screening levels for aquatic effects of perchlorate of 5 mg/L and 0.6 mg/L, respectively.

In their response to comments on the draft toxicological review and risk characterization, the U.S. EPA [28] presented additional data that fulfilled the data requirements for calculating AWQC (see Chapter 5, Table 5.8). The U.S. EPA [28] used these data to calculate acute and chronic AWQC for perchlorate of 22.3 mg/L and 10.3 mg/L, respectively, but cautioned that because these benchmarks "have not been promulgated by the Office of Water nor have undergone full peer review, they cannot be considered national ambient water quality criteria at this time."

Some questions arose regarding the data used to calculate the acute and chronic AWQC. For example, U.S. EPA [28] notes that in the *Pimephales promelas* toxicity test that resulted in an LC_{50} of 614 mg/L, sodium chloride at the same sodium concentration caused toxicity. The reported results may reflect sodium toxicity rather than perchlorate toxicity. Dean et al. [30] calculated AWQC according to the standard protocol using only the relevant acute toxicity test data. Those calculations produced an acute perchlorate AWQC of 20 mg/L and a chronic AWQC of 9.3 mg/L.

In addition to calculating AWQC and similar benchmarks, the U.S. EPA looked at the potential for perchlorate exposure to disrupt endocrine function at particular points in the frog life cycle. In addition to reviewing data from LHAAP (see case study), U.S. EPA [30] reported on unpublished work by Tietge and Deitz that evaluated the effect of perchlorate on the development and metamorphosis of the African clawed frog (*Xenopus laevis*). This work indicated a No Effect concentration of 0.062 mg/L and a Low Effect concentration of 0.25 mg/L. Tietge et al. [30] later published this work. Due to concerns that the chronic AWQC generated using the standard protocol might not be sufficiently protective of sensitive life stages of amphibians, U.S. EPA [30] used the geometric mean of these two values to generate "an interim chronic benchmark of 0.12 mg/L."

6.2 Summary and Conclusions

Regulators use the results of research into toxicological effects and evaluations of potential exposures in the context of other scientific and non-scientific issues to derive enforceable standards and guidance levels for perchlorate. The U.S. EPA has developed screening levels by which to gauge the possible effects of perchlorate contamination on aquatic and terrestrial animals, and both the U.S. EPA and various state agencies have developed drinking water guidelines to protect human health. Table 6.2 summarizes regulatory standards and screening levels for various species, to show the ranges of perchlorate concentrations which may be of concern. (This table was developed from information presented previously in Chapters 5 and 6.)

In February 2005, the U.S. EPA established a final perchlorate oral reference dose of 0.0007 milligrams perchlorate per kilogram body weight per day. This reference dose represents the consensus position on the health effects associated with chronic human exposure to perchlorate. It represents the oral dose that a person, including sensitive subgroups such as an expectant mother, fetus, or newborn, could be exposed to on a daily basis that is not anticipated to cause adverse health effects over a person's lifetime. Using this reference dose, risk managers at the U.S. EPA derived a Drinking Water Equivalent Level of 24.5 μg perchlorate per liter of water, which represents the concentration of perchlorate in drinking water that will have no adverse effect on the general population and sensitive subpopulations. This level incorporates the assumptions that all exposure to perchlorate comes from drinking water and that concentrations above 24.5 μg/L are not necessarily considered unsafe given the margin of safety incorporated into this value. Risk managers within a number of state or federal agencies have derived different drinking water limits or cleanup goals for perchlorate based on different interpretations of the toxicological data, differences in exposure assumptions, and in the context of other relevant scientific, economic, and political issues that inform the risk management decision process.

The U.S. EPA has also evaluated the potential effects of perchlorate on aquatic organisms and calculated acute and chronic AWQC for perchlorate of 22.3 mg/L and 0.12 mg/L, respectively. The effort by the U.S. EPA represented criteria development at two different levels: 1) AWQC calculation to protect most populations of aquatic organisms most of the time, and 2) criteria derivation specifically to protect sensitive life cycle stages (e.g., from endocrine disruption). These criteria are conservatively developed to address potential adverse effects to a sensitive life stage of a sensitive species. Those effects may be balanced, however, by the following observations. As Goleman et al. [31] noted, removal of perchlorate reversed the adverse effects caused by exposure of perchlorate. Further, because perchlorate competitively inhibits iodide uptake, then potentially fewer effects might be observed if iodide is present in sufficient quantities [29]. Most importantly, it is not clear at this time how the effects

Table 6.2 Summary of Regulatory Limits and Screening Levels

Species	Perchlorate Concentration			Notes
	Soil (µg/kg)	Food (µg/kg)	Water (µg/L)	
Human	100 to 5,000 *	-	24.5	U.S. EPA DWEL based on acceptable dose of 0.0007 mg/kg·day
			1-18	Various State guidelines
Raccoon	-	> 5,500,000	> 31,000	No effect observed at those concentrations in one field study
Rat	-	0.01	-	Screening level represents wet weight of plant tissue
Prairie Vole	-	-	-	Acceptable dose 1 mg/kg·day
Earthworm	1,000	-	-	Screening level
Aquatic organisms				
- Acute	-	-	22,300	
- Chronic	-	-	10,300	AWQC, not promulgated
- Chronic, based on endocrine disruption	-	-	120	"Interim chronic benchmark"

* Proposed soil clean-up standards in Massachusetts; applicable value depends on potential exposure. Site-specific cleanup goals ranged to as high as 72 mg/kg.

observed in the laboratory and in the field, such as delayed metamorphosis, translate into adverse effects on populations of aquatic organisms.

6.3 References

[1] U.S. EPA, Drinking Water Contaminant Candidate List, Washington, D.C., Doc NoEPA/600/F-98/002, 1998.

[2] *U.S. EPA*, s.v. "Unregulated Contaminant Monitoring Rule Fact Sheet," http://www.epa.gov/safewater/ucmr/ucmr1/factsheet.html (accessed January 2006).

[3] *U.S. EPA*, s.v. "Integrated Risk Information System [IRIS]," February 12, 2005, http://www.epa.gov/iris/subst/1007.htm#doccar (accessed February 18, 2005).

[4] National Research Council (NRC) of the National Academy of Science (NAS), *Health Implications of Perchlorate Ingestion*, Committee to Assess the Health Implications of Perchlorate Ingestion, National Academies Press, January 2005.

[5] Memorandum from: Susan Parker Bodine, Assistant Administrator, U.S. EPA Office of Solid Waste and Emergency Response, to Regional Administrators, Subject: Assessment Guidance for Perchlorate, January 26, 2006.

[6] U.S. EPA, *Exposure Factors Handbook*, Vol. I-III, EPA/600/P-95/002Fa,b,c, Washington, D.C., 1997.

[7] California Environmental Protection Agency, *Public Health Goal for Perchlorate*, Office of Environmental Health Hazard Assessment (OEHHA), April 2005.

[8] *New Jersey Drinking Water Quality Institute (Department of Environmental Protection)*, s.v. Maximum Contaminant Level Recommendation for Perchlorate, http://www.state.nj.us/dep/watersupply/perchlorate mcl 10 7 05.pdf, October 7, 2005.

[9] *Massachusetts Department of Environmental Protection, Office of Research and Standards*, s.v. Perchlorate Toxicological Profile and Health Assessment, http://www.mass.gov/dep/toxics/perchlor.pdf, May 2004.

[10] Massachusetts Department of Environmental Protection, *Massachusetts Interim Drinking Water Advice for Perchlorate*, Office of Research and Standards, April 16, 2002.

[11] *Massachusetts Department of Environmental Protection*, s.v. Draft Report, The Occurrence and Sources of Perchlorate in Massachusetts, August 2005, http://www.mass.gov/dep/, October 2005.

[12] *U.S. EPA*, s.v. "State Perchlorate Advisory Levels as of April 20, 2005,"

http://www.clu-in.org/contaminantfocus/default.focus/sec/perchlorate/cat/Policy and Guidance, (accessed November 23, 2005). http://www.epa.gov/fedfac/pdf/ stateadvisorylevels.pdf.

[13] *Massachusetts Department of Environmental Protection*, s.v. Notice of Public Hearing: Perchlorate Standards, http://www.mass.gov/dep/, March 10, 2006.

[14] U.S. EPA, *Drinking water criteria document on nitrate/nitrite*, Prepared by Life Systems, Inc., Cleveland, Ohio, for the Criteria and Standards Division, Office of Drinking Water, U.S. Environmental Protection Agency, Washington, D.C., PB-91-142836, 1990.

[15] Howd, R.A., Brown, J.P., and Fan, A.M., Risk Assessment for Chemicals in Drinking Water: Estimation of Relative Source Contribution, Office of Environmental Health Hazard Assessment (OEHHA), California Environmental Protection Agency (Cal/EPA), Oakland, CA, presented as a poster at the 43rd annual meeting of the Society of Toxicology, Baltimore, Maryland, March 21-25, 2004. http://www.oehha.ca.gov/water/reports (accessed December 2005).

[16] *U.S. EPA*, s.v. "Federal Facility and Superfund Sites Where Action Has Been Taken to Address Perchlorate Contamination as of August 2004," http://www.epa. gov/fedfac/documents/perchlorate_site_summaries.htm.

[17] *U.S. EPA*, s.v. "Data from the UCMR 1, covering the period 2001-2005." http://www.epa.gov/safewater/ucmr/data.html#ucmr1 (accessed January 26, 2006).

[18] United States Food and Drug Administration (FDA), Exploratory Data on Perchlorate in Food, Office of Plant and Dairy Foods, November 2004.

[19] Kirk, A.B. et al., Perchlorate in Milk, *Environmental Science and Technology,* 37, 4979-4981, 2003.

[20] Kirk, A.B. et al., Perchlorate and Iodide in Dairy and Breast Milk, *Environmental Science and Technology* 39, 2011-2017, 2005.

[21] *Massachusetts Department of Environmental Protection,* s.v. Proposed Changes to the Massachusetts Contingency Plan-310 CMR 40.0000, http://www.mass.gov/dep/, March 8, 2006.

[22] *U.S. EPA Region III,* s.v. Risk Based Concentration Table, October, 2005, http://www.epa.gov/reg3hwmd/risk/human/index.htm (accessed February 6, 2006).

[23] *Office of Under Secretary of Defense,* s.v. Memorandum regarded Policy on DoD Required Actions Related to Perchlorate, http://www.dodperchlorateinfo.net/, January 26, 2006.

[24] U.S. EPA, *Perchlorate Environmental Contamination: Toxicological Review and Risk Characterization,* NCEA-I-0503, January 16, 2002.

[25] Stephan, C.E. et al., *Guidelines for Deriving Numerical National Water Quality Criteria for the Protection of Aquatic Organisms and Their Uses,* U.S. EPA Office of Research and Development, Environmental Research Laboratories: Duluth, MN; Narragansett, RI; and Corvallis, OR, PB85 227049, 1985.

[26] U.S. EPA, *Final Water Quality Guidance for the Great Lakes System: Final Rule,* 60FR15365, 1995.

[27] U.S. EPA, *Great Lakes Water Quality Initiative Tier II water quality values for protection of aquatic life in ambient water,* Support documents, November 23, 1992.

[28] U.S. EPA, *Disposition of Comments and Recommendations for Revisions to Perchlorate Environmental Contamination: Toxicological Review and Risk Characterization External Review Draft,* dated January 16, 2002, EPA/600/R-03/031, October 2003.

[29] Dean, K.E. et al., Development of freshwater water-quality criteria for perchlorate, *Environmental Toxicology and Chemistry,* 23, No. 6, 1441, 2004.

[30] Tietge, J.E. et al., Metamorphic inhibition of Xenopus laevis by sodium perchlorate: effects on development and thyroid histology, *Environmental Toxicology and Chemistry,* 24, No. 4, 926, 2005.

[31] Goleman, W.L. et al., Environmentally relevant concentrations of ammonium perchlorate inhibit development and metamorphosis in Xenopus laevis, *Environmental Toxicology and Chemistry,* 21, No. 2, 424, 2002.

CHAPTER 7

How Can Perchlorate Be Treated?

Treatment technologies used to effect perchlorate remediation strategies can be classified according to the fate of the perchlorate. In general, treatment may separate contaminants from the medium of interest (soil, groundwater, or sediments), immobilize the contaminants, or destroy the contaminants. The physical and chemical properties of the contaminant molecule determine how it may be separated, immobilized, or destroyed. Due to the solubility of perchlorate salts, perchlorate is typically treated using separation or destruction technologies, but not immobilization techniques. Contaminants that have been separated from the medium require further treatment or disposal.

As a result of the high solubility and mobility of perchlorate, groundwater contamination is a greater concern than soil remediation at many sites. Most discussions of perchlorate remediation focus on groundwater rather than soil. Table 7.1 lists common treatment techniques and indicates their general applicability to soil or groundwater.

Table 7.1 Applicability of Common Treatment Techniques

Form of Treatment	Technology	Applicability	
		Soil	Groundwater
Separation	Ion exchange		♦
	Standard Granular Activated Carbon		♦
	Cationic-substance coated media		♦
	Membrane filtration		♦
	Capacitive deionization/Carbon Aerogel		♦
Destruction	Bioreactors		♦
	In situ biodegradation	♦	♦
	Thermal Destruction	♦	♦
	Phytoremediation	♦	♦

Note: Phytoremediation can also be a separation technology, depending on the biological processes involved.

This chapter describes the most common perchlorate treatment technologies, including the theory behind each technology and information regarding the flexibility and effectiveness of each technology for perchlorate remediation. Technology selection depends on the flow rate or volume of material, geochemistry of the water and/or concentrations of other contaminants, discharge limits or other clean-up goals, and levels of perchlorate. Each technology tends to be effective over a particular range of perchlorate concentrations. For discussion purposes in this chapter, very low concentrations of perchlorate are defined as below 10 µg/L, low concentrations range from 10 to 100 µg/L, moderate concentrations range from 100 to 1,000 µg/L, and high concentrations are above 1,000 µg/L. A summary table of treatment technologies and their suitability for various site characteristics is also presented at the end of the chapter for easy reference.

Definitions
- Bed Volume - One bed volume is the volume of water that can be contained in a specific treatment vessel at any one time. Bed volumes are often calculated based on an empty vessel rather than one filled with a treatment medium, as the volume of the medium is considered to be negligible. This measure is used to compare efficiencies of different treatment media to remove a contaminant from the influent water. Efficient media can treat higher numbers of bed volumes of contaminated water than inefficient media. This is a unit-less measure (e.g., 100,000 bed volumes) as it is dependent on the size of the treatment vessel.
- Breakthrough - Breakthrough is usually defined as the time that contaminants begin to be detected in the effluent of a treatment vessel. It is occasionally defined as the time the contaminant concentration in the effluent reaches the regulatory clean-up limit. Units are in time (e.g., minutes, days, or years).
- Empty Bed Contact Time (EBCT) - EBCT is the optimal time a drop of fluid takes to flow through an empty treatment vessel. Although the treatment vessel is normally filled with a treatment medium, it is assumed that the volume of the medium is negligible, such as in the case of granular activated carbon. Units are normally in minutes.
- Facultative Anaerobes - An organism that can live in the absence as well as in the presence of atmospheric oxygen.
- Fluidized Bed - Particulate fluidization utilizes an upward flow of liquid to fluidize a bed of solid particles. Fluidization is the point at which a packed bed of solids begins to exhibit fluid-like properties, such as movement of the solid particles in the bed and the expansion of the bed. In water treatment, the upward flow of water into the vessel is controlled so that the flow keeps the individual grains of the treatment medium separate but not so forcefully that the medium flows out of the top of the vessel.
- Regeneration - When the capacity of a treatment medium is exhausted and no more contaminants can be adsorbed to it, the medium may in some cases be regenerated. In the case of ion exchange resins, the resin in the treatment vessel is flushed free of the contaminant ions and contacted with a solution of replacement ions. In the case of granular activated carbon, the carbon can be removed from the vessel and heated to destroy the contaminants, then replaced into the vessel.

- **Service Flow Rate** - The optimal water flow through a treatment system, usually used for ion exchange resins. Units are in time, but this is misleading as it really means the flow (e.g., gallons per minute) of water per cubic foot of the medium within the treatment vessel.

7.1 Groundwater Extraction

Several of the treatment technologies described in this chapter apply to groundwater pumped from a contaminated aquifer. Engineers apply other technologies *in situ*. To provide a basis for discussing the treatment of extracted groundwater, this section briefly describes techniques for extracting groundwater. It is adapted from *Fundamentals of Hazardous Waste Site Remediation* [1], with permission.

Groundwater extraction and treatment, commonly called pump and treat, can be used to remove contaminant mass and restore an aquifer to clean-up goals, or to manage contaminant migration by controlling hydraulic gradients and thus the spread of contamination. Groundwater is commonly pumped from extraction wells, well points, or an extraction trench. Each of these extraction methods is discussed below.

7.1.1   Extraction Wells

The design of an extraction well system depends on the aquifer characteristics, purpose of the system (gradient control vs. aquifer restoration), and the size of the plume. The aquifer characteristics determine how much water can and should be pumped from each well. The siting of the wells and, as a result, the total quantity of water to be pumped, depends on the purpose of the system and the size of the plume.

7.1.1.1   Well Construction and Pumps

A typical extraction well in an aquifer is oriented vertically, although horizontal wells are occasionally used. The major components of a typical well are the casing, screen, filter pack and seal, and pump. Groundwater flows into the well through the screen, which is commonly made of slotted polyvinyl chloride (PVC) or stainless steel pipe or continuous-slot shaped steel wire. The casing is sized to accommodate the pump and is typically four inches in diameter or larger. The filter pack, typically coarse to medium sand or fine gravel, holds the borehole open, prevents excessive soils from entering the well, and allows water to flow freely into the well. The pump provides the mechanical energy needed to draw water into the well and convey it up to the ground surface and through the header (piping) to the treatment plant. Two types of pumps are commonly used to extract groundwater: pneumatic pumps and electric submersible pumps. Pumps are sized based on the energy (head) required, which is calculated from the amount of water to be pumped and the resulting velocity through the piping; the projected friction losses through the piping; the vertical distance (elevation head) to which the water must be lifted out of the well; and the efficiency of the pump.

The natural constituents of groundwater can foul or damage the well screen or pump. When corrosive compounds are present, the design must incorporate materials that are not susceptible to corrosion. Free carbon dioxide (CO_2) at concentrations greater than 10 to 15 mg/L can corrode steel and cast iron. Hydrogen sulfide (H_2S), encountered in tidal estuaries and ocean-front marshes, corrodes steel, cast iron, brass, and bronze at concentrations over 1 to 3 mg/L. Chloride ions corrode stainless steel, particularly at levels above 500 mg/L. Certain compounds can precipitate on well screens, sand packs, piping, and pumps; at high concentrations, these precipitates restrict flow and can damage pumps. At concentrations over 2 to 3 mg/L in groundwater, iron deposits can become a severe problem, either from precipitates of iron salts or from an iron-laden slime produced by iron-fixing bacteria. Some precipitation may occur at concentrations as low as 0.25 mg/L. Hardness over 200 mg/L can also cause problems from encrustation with calcium and magnesium carbonates. When fouling occurs, precipitates are removed by treating the well with a dilute solution of inhibited hydrochloric acid or another acid. Chlorine and hypochlorite are used to kill iron-fixing bacteria.

7.1.1.2 Modeling the Flow to a Single Extraction Well

Pumping induces groundwater flow to a well, imposing an artificial hydraulic gradient and creating a cone of depression in the water level around the well. The decrease in the water table or head caused by pumping is called drawdown. The pumping rate and the radius of influence of each well can be estimated by mathematical models of groundwater behavior and/or by pumping tests. Two- and three-dimensional simulations of the effects of a network of monitoring wells require computer models. "Real world" systems contain many complicated variables that cannot be readily modeled. Simplifying assumptions must be built into groundwater models in order to make the models manageable. In general, models of the flow from an extraction well assume equilibrium conditions and assume ideal, homogenous aquifer conditions. As a result, they only approximate the performance of a pumping system under actual (non-equilibrium, non-ideal, non-homogeneous) conditions. Developing a practical design from model projections requires considerable professional judgment.

7.1.1.3 Siting Extraction Wells

Groundwater remediation usually requires a network of extraction wells rather than a single well. The location of the wells and the total amount of water to be pumped depends on the goal of remediation in addition to aquifer characteristics. This goal may be to restore aquifer conditions, to control the groundwater gradient and prevent a plume from expanding, or both.

- Aquifer restoration requires the removal of the contaminant mass dissolved in the groundwater and adsorbed to saturated soils. Extraction wells are placed throughout the plume to withdraw contaminated groundwater and induce flushing with clean water. These wells produce a stream of ground-water for treatment that contains relatively high concentrations of contaminants.

- Gradient control requires a system of extraction wells at or near the leading edge of a plume to prevent contaminated groundwater from migrating further. Because gradient control plumes are pumping relatively clean water, the costs of treating the water can be lower than the costs of treating water from wells in the center of a plume, depending on the flow rate of water to be treated.
- At the periphery of a plume, wells are sited for migration control to minimize - to the extent possible - the volume of clean water drawn into the well. Higher flow rates require higher costs for treating the water.

Thus, the goal of remediation determines the placement of the well network within a plume, at the leading edge, or both. Wells are sited so that the radius of influence of adjacent wells overlap and no flow occurs between the wells. Computer models are often used to simulate the flow patterns induced by different well configurations and pumping rates in order to optimize the design.

7.1.2   Well Point Systems

Groundwater can be extracted from a system of well points through a header using a common vacuum pump mounted at the surface, rather than using a water pump in each well. The pump lifts the water from the well points and draws it through the header by suction. A valve on the hose between the header and each well point regulates the flow from each well point. Adjusting the flow from different well points is called tuning. A well point system differs from an extraction well network in the use of a common vacuum pump, rather than a pump in each well, and in the construction of the wells. Because of the dependency on vacuum, well point systems are limited to a suction lift of about 15 ft, with a practical maximum of 28 ft.

A well point is a screen made of wire mesh, slotted plastic, or other material. It can be installed by being driven into the ground, by jetting (using a high-velocity stream of water to cut the borehole), and into drilled boreholes, similar to an extraction well. Filter sand is placed around a well screen installed by jetting or drilling.

Well point systems are used less commonly than extraction well networks in long-term groundwater remediation systems. All else being equal, a well point system costs less than a network of extraction wells with a pump in each well. However, well point systems have several limitations. They are limited to shallower depths and lower pumping capacities than extraction well networks. Well points can be particularly susceptible to fouling, depending on the operation and the site geochemistry. If the well point screen is only slightly below the water table, air will be drawn into the system with the groundwater. Aeration can cause some ions, such as iron, to precipitate, and can stimulate the growth of aerobic bacteria. The induced vacuum can also enhance the precipitation of certain compounds, such as calcium bicarbonate or certain iron carbonates.

7.1.3   Interceptor Trenches

Contaminated groundwater is sometimes recovered from an interceptor trench or subsurface drain. In permeable soils, the trench is ideally excavated down to an impermeable layer, to prevent contaminated groundwater from flowing under the trench. A perforated pipe is wrapped in filter fabric and lowered into a trench. The trench is then backfilled with gravel or crushed stone. The backfill is more permeable than the surrounding soil and therefore transmits water more readily. The pipe conveys groundwater to a sump, where the groundwater is pumped to a treatment system.

This basic design has many variations, most commonly:

- A french drain does not contain a pipe to convey the water; the water is simply transmitted through the gravel in the trench.
- The downgradient side of the trench may be lined with a geomembrane (e.g., high density polyethylene or HDPE) to prevent inflow of water from the downgradient side.
- A plastic drainage material may be placed in the trench rather than gravel backfill. Sheets of corrugated plastic are placed vertically in the trench to provide drainage.

An interceptor trench is oriented perpendicular to the direction of groundwater flow to prevent the migration of contaminated groundwater. It functions like an infinite line of extraction wells that lowers the water table to the depth of the drain. Interceptor trenches are used in relatively low permeability soils, where wells or well points would have to be spaced quite closely. Trench construction has several limitations that do not apply to installation of extraction wells:

- Construction costs generally limit trench construction to a depth of 40 feet or less using conventional equipment, although deeper trenches have been built.
- Underground utilities are common on industrial or urban sites. It is more difficult to construct a trench under utilities or to reroute utilities away from a trench than it is to site wells away from utility lines.
- Interceptor trenches can also be more difficult to maintain than extraction wells, particularly where natural groundwater constituents such as iron or hardness are high enough to cause fouling.

Once pumped from the aquifer, whether via extraction wells, well points, or an interceptor trench, groundwater must be treated to remove perchlorate. Section 7.2 describes techniques to separate perchlorate from water. Section 7.3 describes technologies that can destroy perchlorate in water or soil.

7.2 Separation Technologies

This section describes five techniques for separating perchlorate from extracted groundwater: ion exchange, granular activated carbon, cationic-substance coated media, membrane filtration, and capacitive deionization using carbon aerogel (an air-filled carbon gel).

7.2.1   Ion Exchange

Ion exchange (IX) is one of the most common methods to treat perchlorate in groundwater. It is easy to implement, can remove a wide range of concentrations of perchlorate from groundwater, and has the longest track record of the perchlorate remediation technologies.

7.2.1.1   Background and Theory

Ion exchange is a physical-chemical process in which charged functional groups on the surface of a solid attract and thereby remove ions from water via electrostatic forces. The solids used in ion exchange media are polymer resins with cross-linking (i.e., connections between long carbon chains in a polymer) that form charged functional groups. To increase the surface area and thereby maximize the number of these active sites, manufacturers usually cast ion-exchange resins in the form of small porous beads, generally on the order of 250 to 425 microns (ASTM sieve mesh sizes #60 to #40).

In the case of perchlorate IX resins, the functional groups have positive charges. The resins are typically initially dosed with an innocuous anion such as the chloride ion, which attaches to these positively charged functional groups. The perchlorate anion in the influent water is also attracted to the positively charged functional groups, and because it has stronger attractive force, it will replace the chloride ion [2]. Ion exchange thus removes the perchlorate ion from the aqueous phase by "exchange" with the chloride ion, populating the treated water with chloride ions. "Breakthrough" occurs when the exchange sites can no longer sorb perchlorate or the perchlorate on theses sites begins to desorb. Once the resulting perchlorate concentration in the effluent reaches a trigger point (which may be a regulatory limit or simply the detection limit for the site), the vessel bed has experienced breakthrough. At that point, operators must replace or regenerate the IX resins.

Extensive research has focused on developing IX resins that would selectively remove perchlorate when the water contains other competing anions, notably chloride, sulfate, nitrate, and bicarbonate, at higher concentrations. Strong-base anion-exchange resins have proven to be very effective in removing perchlorate from water [3]. The following equation describes the removal of perchlorate from water by strong-base anion exchange resins:

$$\text{Resin-Cl}^- + \text{ClO}_4^- \ll \text{Resin-ClO}_4^- + \text{Cl}^- \qquad (7\text{-}1)$$

Once breakthrough occurs, the resin can potentially be regenerated with sodium chloride (NaCl) according to the reaction below, although the process requires a catalyst or heat:

$$Resin\text{-}ClO_4^- + NaCl \ll Resin\text{-}Cl^- + Na^+ + ClO_4^- \qquad (7\text{-}2)$$

A paper prepared for the journal *Water Conditioning and Purification* describes four types of IX resins, including Type I styrenic strong-base resins, acrylic strong-base resins, nitrate-selective resins, and perchlorate-selective resins [4]. Each type of resin is discussed briefly below.

- Acrylic strong base resins have a relatively low affinity for perchlorate. Thus acrylic strong-base resins do not remove perchlorate efficiently, but they have been used in the past because they can be easily regenerated.
- Type I styrenic base resins have a higher capacity for perchlorate removal than acrylic strong base resins. It is often used as the standard for IX resin performance comparisons. It has been commonly used for the removal of nitrates from groundwater, but it is not as useful for perchlorate removal in locations where perchlorate concentrations are moderate to high (100 µg/L or higher).
- Nitrate selective resins have close to two times the removal efficiencies of the Type I styrenic resins. These resins are not, however, considered to be regenerable.
- Perchlorate-selective resins remove perchlorate much more efficiently than the Type I styrenic resins and four to five times more effective than nitrate selective resins. Perchlorate-selective resins' strong affinity for perchlorate and much lower affinity for other competing anions allows a treatment vessel to remain on-line for much longer than other resins for removal of perchlorate from groundwater. Perchlorate-selective resins are highly effective for treatment of groundwater containing concentrations of perchlorate on the order of 100 µg/L or higher.

Table 7.2 presents a summary of information on the four types of resins [4]. The Separation Factor denotes the relative removal efficiency of the resin, using the exchange rate between perchlorate and the chloride ion for illustration purposes.

Table 7.2 Comparison of Ion Exchange Resins [4]

Ion Exchange Resin Type	Separation Factor ($\alpha\text{-}ClO_4/Cl$)	Capacity (bed volume)	Cost Comparison (X multiplier per cubic foot of resin)	Cost Comparison (factored for Capacity)
Acrylic Strong Base Resin	4-5	300	1X	30X
Type I Styrenic Resin	100-150	4,000	1X	2.5X
Nitrate Selective Resin	>200	7,000	2X	2.5X
Perchlorate Selective Resin	>1,000	50,000	3X	1X

Note: Capacity is based on typical California water, with 50 µg/L perchlorate, 44 mg/L sulfate, 40 mg/L nitrate, 170 mg/L bicarbonate, and 13 mg/L chloride.

Until recently, the cost of various non-regenerable IX resins such as the perchlorate-selective resin was considered prohibitive for treatment of low concentrations of perchlorate. However, costs decreased in the last two years by fifty percent due to competitive market factors. These resins would likely be appropriate for treatment of low concentrations of perchlorate. Please note that the cost comparisons are valid as of the time this book was written. In the competitive marketplace, prices of the different types of resins may vary significantly over time. Comparisons shown in Table 7.2 are therefore approximate.

7.2.1.2   Application

Typically, ion-exchange resins are placed in columns, or vertical cylindrical pressure vessels. As the water flows downward through the column, ions in the water exchange with ions on the resin. The column is sized based on the flow rate, the capacity of the resin under site-specific conditions, and the intended frequency of regeneration. The ion-exchange resin is supported on a screen to prevent the loss of the resin beads. Influent water is distributed across the top of the resin bed.

Figure 7.1 Typical Ion Exchange Process
(Courtesy AMEC Earth and Environmental, Inc.)

Engineers typically size ion exchange vessels based on the service flow rate, which is the optimal water flow through the system in units of cubic feet of ion exchange resin in the treatment vessel. Typical service flow rates are 1 to 4 gallons per minute (gpm) per cubic foot, or 10 to 40 bed volumes per hour [5].

Ion Exchange vs. Carbon Adsorption

In IX, anions such as perchlorate adhere to resins based on their ionic charge. Granular activated carbon (GAC) relies on weaker Van der Waals forces to adsorb complex molecules. The reaction time for an ion exchange mechanism is therefore quicker than for GAC systems.

The service flow rate for an IX column is the industry's equivalent of the empty bed contact time (EBCT) used in GAC systems, where the EBCT represents the residence time of fluid flowing through an empty treatment vessel. The service flow rates for ion exchange systems typically range from 1 to 4 gallons per minute per cubic foot, compared with EBCTs of 5 to 20 minutes for GAC systems. This difference allows for a greater volume of water to be treated by the same size of vessel.

The unit may be plumbed to allow for regeneration through either downward flow or upward flow of a regenerant solution [6]. When the ion-exchange capacity of the resin is exhausted, the column may in some cases be backwashed to remove trapped solids and then regenerated. The resin is flushed free of the newly exchanged ions and contacted with a solution of replacement ions. Regeneration is initiated after most of the active sites have been used and the ion exchange is no longer effective. Regeneration thus allows for the reuse of resin beads, although the process is only effective for resins that have a relatively low affinity for perchlorate such as the acrylic resins.

Regenerant solutions include 12% brine solution of sodium chloride, which can recapture approximately 7% of the perchlorate from the resin, and ferric chloride in solution with hydrochloric acid, which can recapture close to 100% of the perchlorate from the resin [7]. The regeneration process produces a highly acidic waste stream that contains concentrated contamination and which generally requires disposal or further treatment. Treatment processes under study include biological destruction [8] and thermal destruction of the waste brine [9]. In addition, perchlorate in the regenerant brine may be destroyed by a catalytic process using hydrogen. The chemical reaction produces sodium chloride and water [10].

Perchlorate-selective resins cannot be backwashed because not enough perchlorate is removed from the resin to make it cost effective. The alternative to regeneration for such resins is to simply replace the resin with fresh material and dispose of the spent resin.

Operating concerns include [1]:

- Treatment/disposal of concentrated regenerant and/or disposal of ion exchange resin
- Fouling due to high levels of suspended solids in the water (> 10 mg/L), which will clog or blind the resin
- Bed compaction/biofouling, which may require the resin to be replaced before it reaches its theoretical bed life
- The concentrations of related anions in the influent stream that may compete with perchlorate for attachment to the positively charged functional groups, such as nitrate, sulfate, bicarbonate, and chloride
- The concentration of exchanged ions in the treated water

Many groundwater treatment systems, at sites in California and elsewhere, incorporate ion exchange. Table 7.3 lists some of the sites where ion exchange has been used to treat perchlorate, alone or as one of several technologies [11] [12].

Table 7.3 Perchlorate Treatment Systems Using Ion Exchange Resins [11] [12]

Site	Perchlorate Concentration ($\mu g/L$)	Ion Exchange Resin Used
Edwards Air Force Base Antelope Valley, CA	160,000	Purolite A-530E selective, regenerable bi-functional resin (perchlorate selective resin, a.k.a. Oak Ridge D-3696™)
Lawrence Livermore National Laboratory - Livermore, CA	20	Sybron IONAC® SR-7 (nitrate selective resin)
San Gabriel Valley, CA (Azusa, La Puente, Baldwin Park)	600	Calgon Carbon CalRes2000 SBA Type 1 (acrylic strong base resin)
Aerojet, Sacramento, CA	50	Rohm and Haas Amberlite® PWA2 (perchlorate selective resin)
Kerr-McGee Henderson, NV	350,000	Calgon Carbon ISEP™ used until 2004, replaced with biological fluidized bed reactor
Massachusetts Military Reservation Cape Cod, MA	40	Purolite A520E (nitrate selective resin)

Case Study: San Gabriel Valley Superfund Site/Baldwin Park, La Puente, California

In California, concentrations of perchlorate in groundwater tend to range from 10 to 100,000 µg/L [4], in the presence of high concentrations of competing anions, as tabulated below. At La Puente, as noted previously, the perchlorate concentrations range from 40 to 160 µg/L.

Average Geochemistry in California Groundwater [4]	
Analyte	Concentration
Perchlorate (μg/L)	50
Chloride (mg/L)	13
Nitrate (mg/L)	40
Sulfate (mg/L)	44
Bicarbonate (mg/L)	170

When chemists first detected perchlorate in La Puente's groundwater in 1997, few treatment options for perchlorate in groundwater existed. The only ion exchange resins available were the Acrylic Strong Base and Type 1 Styrenic anion exchange resins. Biological treatment systems were just being developed. *In situ* methods were not feasible given the size of the groundwater plume, the relatively low concentrations in groundwater, and most importantly, because groundwater at La Puente provides the municipal water supply, the costs associated with extraction have already been borne for this purpose.

ISEP™ Carousel for Perchlorate Treatment at La Puente, California [13]

La Puente instituted ion exchange treatment of the groundwater. Construction was completed in March 2002, and the plant was in operation by the Summer of 2002. Because of the flow rate that the district required and the limited ability of the available ion exchange resins to remove perchlorate, the district selected the Ion SEParator™ (ISEP) process engineered by Advanced Separation Technologies Incorporated, a subsidiary of the Calgon Carbon Corporation. In this remediation process, a distributor on a rotating carousel directs all incoming and out-going fluid streams into one of 10 or more treatment vessels containing Purolite A850 (Acrylic Strong Base Type 1) ion exchange resins in the carousel. Each treatment vessel cycles through two phases. In the first phase (adsorption) the resin removes perchlorate from the groundwater. In the second phase (rinse and regeneration) a counter-current flow of rinse-water rinses the vessel and then briny water (7%) is added to re-exchange the perchlorate with chloride, to be ready for a new adsorption phase [14].

What happens to the regenerant solution? Prior to 2004, some treatment facilities discharged the waste to the ocean through a long discharge pipe. However, California regulations recently required treatment of the waste brine. At La Puente, the waste brine is discharged to a dedicated brine disposal line attached to a Perchlorate/Nitrate Destruction Module using catalytic reduction to destroy the perchlorate at a cost of $7 per acre-foot of treated water. An alternative that has been piloted at La Puente is the biological destruction technique mentioned above. The brine is diluted to about 1-3% salt concentration, at which point bioremediation is possible in either a fixed bed or fluidized bed system (described further in Section 7.3).

More recently, another treatment technology has been studied at La Puente. In 2002-2003 the plant performed a pilot test of a hollow fiber membrane biofilm reactor (described in Section 7.3). The results showed that the reactor destroyed both perchlorate and nitrate in a 1.5 gpm system.

7.2.2   Standard Granular Activated Carbon (GAC)

Granular activated carbon (GAC) has not historically been considered an effective treatment technology for perchlorate remediation. However, recent research and implementation has shown that in some situations, GAC can effectively remove perchlorate from groundwater. In addition, GAC is one of the oldest known water treatment technologies and its processes are generally well understood; therefore it is easy to implement a GAC system.

7.2.2.1   Background and Theory

Manufacturers produce GAC by crushing a high-carbon material such as coal, wood, coconut shells or nutshells and then roasting it to make charcoal. A second roasting step, in the presence of steam, creates highly porous granules. These pores provide the extremely high surface area that makes GAC an effective adsorbent. GAC is made in two forms: powdered and granular. In general, the larger the surface area, the more sites available for sorption of contaminants of concern and thus the higher the sorption capacity of the GAC [15].

The mechanism by which perchlorate sorbs onto standard GAC is not well understood. Conceptually, perchlorate interacts with the positively charged surfaces of the GAC particles rather than adsorbing to the inner surfaces of pores in the GAC as would a large organic molecule [16]. GAC has comparatively fewer such charged sites than ion exchange resins. In addition, if the water stream contains other anions such as nitrates, sulfates, chloride, and bicarbonates in significant concentrations, those anions limit GAC's ability to sorb perchlorate ions. GAC beds exhausted from perchlorate adsorption have been replaced rather than regenerated to date, but this practice could potentially change as the mechanisms for perchlorate adsorption become better understood.

7.2.2.2 Application

GAC is typically packed in a vessel with a flow-through column designed to operate under pressure. Typically, contaminated water flows downward through the column. As the water flows through the vessel, contaminants adsorb onto the GAC. Two or three GAC vessels in series are frequently used to remove contaminants from groundwater, as shown in Figure 7.2. Multiple vessels may be used in parallel to provide hydraulic capacity.

GAC vessels are sized to provide sufficient time for the contaminated groundwater to contact the GAC and contaminants to adsorb to the GAC. The sizing of GAC vessels and the design of GAC systems are typically based on empty bed contact time (EBCT), which is the residence time of fluid flowing through an empty vessel assuming that the volume of the GAC is negligible. EBCT values typically range from 5 to 20 minutes per unit. The size is also based on the necessary bed depth; room needed for expansion during backwashing (discussed below); and the bed life, or frequency of GAC replacement. Liquid-phase GAC vessels are available from many vendors in a variety of sizes.

Solids in the influent gradually accumulate on the GAC bed, causing a pressure drop across the GAC. Engineers design large GAC adsorption units intended for long-term use to permit periodic backwashing to remove particulates when the pressure drop becomes too high. Backwashing is accomplished by reversing the flow through the GAC unit, using clean water to expand the GAC bed and remove the solids. Backwashing requires that treatment of contaminated groundwater in that unit must stop or be diverted to the other vessels. The wastewater generated from backwashing should be treated.

When the GAC's adsorption capacity is used up, or spent, contaminant breakthrough occurs. Breakthrough has been variously defined as the time that contaminants begin to be detected in the effluent from the GAC bed, or the time the contaminant concentration in the effluent reaches the clean-up limit. Engineers typically specify two or three columns in series to treat groundwater to prevent the discharge of unacceptable levels of contaminants when breakthrough occurs. In a system using two columns in series, when contaminants break through the first unit (Column 1), the second unit (Column 2) takes the burden of adsorbing the contaminants. At that time, the GAC in the first unit (Column 1) is replaced or regenerated to provide fresh GAC. The flow sequence is then reversed, so that the water flows first through Column 2, then through Column 1, which is now the cleanest GAC bed.

Spent GAC may be landfilled or regenerated. During regeneration, the GAC is thermally treated to remove and destroy adsorbed organic contaminants. As noted above, however, spent carbon from perchlorate treatment is not now regenerated.

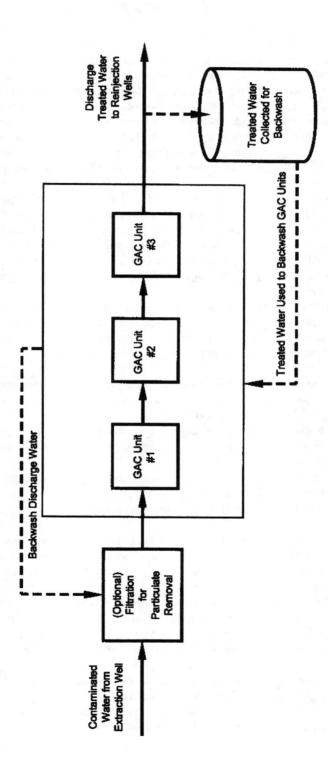

Figure 7.2 Typical Granular Activated Carbon Process
(Courtesy AMEC Earth and Environmental, Inc.)

The potential performance of GAC in a specific application can be predicted by performing bench-scale tests or by analogy to use at a similar site. Bench-scale tests include equilibrium tests and continuous flow GAC tests. Equilibrium tests typically are conducted at constant temperature and are thus referred to as isotherm tests. Isotherm curves provide information on adsorption processes and the extent of surface coverage by the adsorbate. Isotherm curves provide a quick method for comparing the effectiveness of different GACs for various contaminants. The capacity of GAC to adsorb a contaminant varies with the concentration of that compound in solution; adsorption isotherms describe this variation at a constant temperature.

Continuous flow GAC tests are also referred to as rapid small scale column tests (RSSCTs). These tests are performed by using scaled-down columns packed with finer mesh grain sizes of GAC. RSSCTs performed over several days in the laboratory can simulate several months of full-scale operation. Test results provide understanding of the dynamic effectiveness of GAC, and are used to accurately estimate the effective bed life and the optimal EBCT of the GAC, design parameters, and filter bed depth.

GAC units can foul as a result of bacterial growth or precipitation of naturally occurring metals in groundwater. They can also foul or lose capacity due to naturally occurring organics or contaminants other than the target compounds. The following conditions can lead to fouling problems [1].

- Bacteria may grow on the GAC bed if the groundwater contains high levels of biodegradable organics, greater than 20 mg/L [17]. This biological growth can clog the GAC bed.
- Suspended solids > 50 mg/L require backwashing so frequently as to be impractical. Pretreatment (e.g., by sedimentation or filtration) would be required.
- Iron and manganese may precipitate and clog the GAC unit. Pretreatment may be required to remove iron and manganese if their total concentration exceeds approximately 10 milligrams per liter.

Several systems have used standard GAC to remove perchlorate, including a pilot-scale public water supply treatment system in Redlands California [18] and two full-scale groundwater treatment systems at the Massachusetts Military Reservation [19] [20].

The Redlands pilot-scale system successfully reduced perchlorate concentrations from an average of 75 μg/L to less than 4 μg/L (the applicable drinking water standard in California at the time) utilizing an EBCT of 60 minutes. The treatment system processed approximately 1,150 bed volumes until perchlorate breakthrough.

The long EBCT and the low number of bed volumes processed before breakthrough suggest that standard GAC was not very effective for perchlorate removal at this site. The results likely reflect the relatively low influent concentration of perchlorate, and perhaps reflect the high concentrations of anions such as nitrates, sulfates, and bicarbonates that compete for the same sites on the GAC as perchlorate (see the data tabulated below).

Case Study: Massachusetts Military Reservation, Cape Cod, Massachusetts

By 2004, the plume of groundwater contamination at Demo 1 extended for approximately 9,200 ft downgradient, had spread laterally to approximately 1,400 ft and was approximately 90 ft thick. The U.S. EPA required Rapid Response Actions to begin to remove dissolved contaminant mass from groundwater while continuing to evaluate the feasibility of comprehensive remedial actions. In response, the Army installed two pump and treat systems. One, known as the Frank Perkins Road Extraction Treatment and Removal (FPR-ETR) system, lies within the area with the highest levels of contamination and remediates both explosive compounds such as RDX and perchlorate. The second system, nearer the leading edge of the plume and designated the Pew Road Extraction Treatment and Removal (PR-ETR) system, treats primarily perchlorate. These systems extract, treat, and reinject groundwater while the selection, design and construction of a comprehensive solution are under way.

The design began with groundwater modeling to site groundwater extraction and reinjection wells. The design incorporated reinjection of treated water to aid in hydraulic control of the plume. Using the programs MODFLOW and MODPATH coupled with a particle-tracking algorithm, hydrogeologists tested multiple configurations of well locations and pumping rates to determine a design that could capture the plume. In addition to hydrogeologic considerations, concerns that construction would cause widespread ecological damage further constrained the well locations. The positions of wells, trenching and treatment buildings were limited to areas that had already been disturbed. Thus the extraction wells for the Frank Perkins Road system were sited on a tank track to the east of Frank Perkins Road, and the extraction well for the Pew Road system was sited adjacent to a monitoring well on Pew Road.

The decision to reinject treated water into a Sole Source drinking water aquifer imposed stringent requirements for treating water. The relevant standard in force at the time was 1.5 μg/L perchlorate. Thus as a consequence of MMR's location, the design team needed to develop unique treatment systems: the influent would contain perchlorate at concentrations lower than the effluent from groundwater treatment plants at most other sites, and the effluent could only contain extremely low levels of perchlorate, just above the analytical detection level.

Several other challenging conditions were established for the treatment systems. First, the RRA had to meet a tight schedule: twelve months from beginning the design to installing and starting up the extraction, treatment and reinjection systems. Second, the treatment systems had to be mobile and to work in remote locations (although power could be supplied via a combination of overhead and subsurface lines). Finally, the treatment systems also had to provide flexibility for treatment of multiple contaminants. These design constraints combined to limit the allowable size of the treatment vessels, which resulted in a shorter EBCT than what is typically used for RDX and perchlorate treatment.

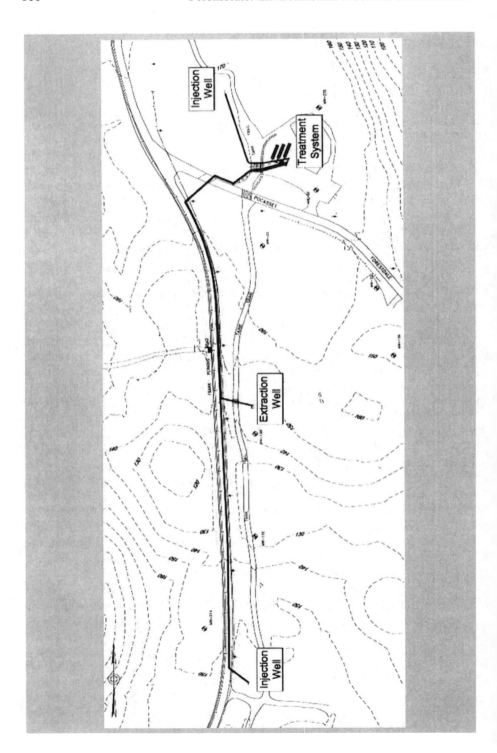

Treatment with this shortened EBCT is feasible due to the low levels of RDX, perchlorate, and other geochemical constituents present in groundwater at the site. The systems were configured as a paired series of three carbon vessels (or a total of six vessels per mobile treatment unit) that would each contain the equivalent of 1,000 pounds of standard GAC. These vessels could then be used to house various IX resins, standard GAC, or even tailored GAC (as described in Section 7.2.3) in the future in any configuration of the treatment media. Due to the different requirements for treating groundwater in each of the systems, each system was configured differently.

Frank Perkins Road Treatment System, Camp Edwards MMR [21]

FPR-ETR System Operations: Sited to extract groundwater from the midpoint of the plume, the FPR-ETR system was designed to treat 220 gallons per minute (gpm) of groundwater containing 1 μg/L RDX and 30 to 35 μg/L perchlorate initially, falling to 15 μg/L perchlorate within a year. Laboratory and field scale studies indicated that the most cost-effective treatment for the system would be standard GAC to remove the RDX, followed by a nitrate-selective ion exchange resin, followed by a polishing GAC vessel, with an EBCT of approximately 10 minutes. With a predicted bed life of 2 months and a consequent change-out frequency of six times per year, standard GAC was considered to be less cost effective as the primary treatment medium for perchlorate than IX resins at that time. Furthermore, in addition to providing RDX removal, the initial vessels in the treatment train were filled with standard GAC to provide additional information regarding GAC's ability to remove perchlorate.

Perchlorate influent concentrations were consistent with the predicted concentrations, with initial concentrations slightly higher than 35 μg/L as shown in the graph of influent and effluent concentrations for the system. Concentrations fell to less than 25 μg/L within 3 months of operation to the end of December 2004, and were between 10 and 15 μg/L during the first 6 months in 2005. RDX concentrations averaged 5 μg/L during the first 3 months. Although the system effectively treated the contaminants at these levels, perchlorate breakthrough was observed sooner than predicted. Breakthrough was observed at 9,000 bed volumes after 1.5 months of treatment using GAC, slightly earlier than the predicted 2 months.

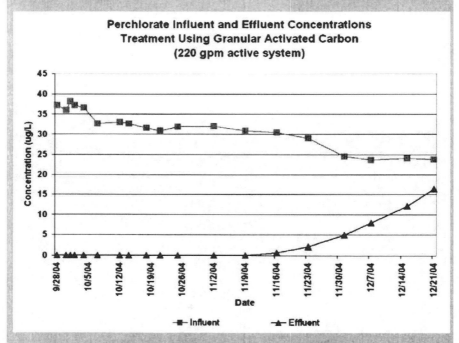

**Perchlorate Influent and Effluent Concentrations
Treatment Using Granular Activated Carbon
(220 gpm active system)**

Breakthrough in First Vessel of MMR FPR-ETR System
at 9,000 Bed Volumes [21]

PR-ETR System Operations: The PR-ETR system, located at the toe of the plume, treated more dilute concentrations of perchlorate in groundwater than the FPR-ETR system. The design was based on influent groundwater concentrations of 1 μg/L RDX and 3 μg/L perchlorate, with an influent flow of 100 gpm. Based on the results of treatability tests, the system would utilize standard GAC with an EBCT of approximately 7 minutes. With a predicted bed life of 4.5 months and a consequent change-out frequency of three times per year, this was considered to be more cost effective than IX resins at that time.

Perchlorate concentrations began to rise approximately 1 month after startup, increasing to a level of 15 μg/L after 3 months of operation, exceeding predictions.

Pew Road Treatment System, Camp Edwards MMR

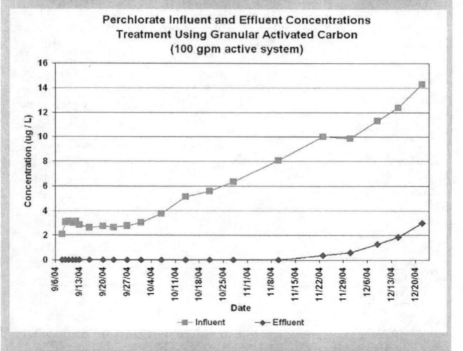

Breakthrough in First Vessel of MMR PR-ETR System
at 17,000 Bed Volumes [21]

RDX concentrations rose to a level of 3 μg/L over the same time period. Although the system was able to treat the contaminants at these levels, perchlorate breakthrough occurred sooner than predicted, at 17,000 bed volumes after 2.5 months of operation.

Why were the FPR-ETR system influent concentration predictions better than those for the PR-ETR system? More groundwater monitoring wells were in place to delineate the plume and thereby provide information sufficient for good model predictions for the FPR-ETR System. As a result, the predicted influent concentrations were likely to be more accurate. Concerns over disturbing the ecosystem by constructing access roads limited the number of monitoring wells upgradient of the PR-ETR system. As a result, there was an information gap of two to three thousand feet upgradient of the extraction well and therefore not enough information was available to anticipate higher perchlorate concentrations. The concentrations in the wells near the system consistently measured between 1 and 3 μg/L for two years' worth of monitoring prior to installation of the system. It was only when the treatment system was in place for a month that the increases were observed.

How do these results compare to data from other sites? The operating data show that the MMR treatment systems remove perchlorate more effectively than the Redlands pilot system, in that the systems can treat more bed volumes of groundwater before breakthrough occurs. This is due largely to the lower influent perchlorate concentrations, especially in the PR-ETR system. It may also be due in part to the lower concentrations of the competing anions in the groundwater; however, the relative sorption capacities of perchlorate with other anions in GAC systems is not well understood.

Perchlorate Treatment by Activated Carbon at Redlands CA and Massachusetts Military Reservation (MMR)

	Redlands	MMR FP-ETR[1]	MMR PR-ETR[1]
Contaminants of Concern (influent)			
Perchlorate (μg/L)	75	33	5
RDX (μg/L)	ND	7	1
Geochemistry			
Chloride (mg/L)	NA	8.9	8.8
Nitrate (mg/L)	5.8	0.17	0.05
Sulfate (mg/L)	30	5.8	4.4
Bicarbonate (mg/L)	145	12.7	12.4
System Characteristics			
Flow Rate (gpm)	Pilot test	220	100
Bed Volumes to Breakthrough	1,150	9,700	21,000
Empty Bed Contact Time (min)	60	7	5

[1]First three months of operations [21] [22]

ND = not detected; NA = not available

The GAC in the FPR-ETR system is not as effective as it is in the PR-ETR system due to the higher concentration of perchlorate in the influent water. Therefore, the FPR-ETR treatment system also includes a treatment vessel containing nitrate selective ion exchange resin to remove perchlorate, while the GAC vessels remain in the system to remove explosives such as RDX. After 16 months of operation a decision was made to also replace spent GAC in the lead vessel in the PR-ETR system with ion exchange resin (the same resin used in the FPR-ETR system), as the influent perchlorate concentrations remain higher than 10 μg/L, and change-outs will be significantly less frequent, on the order of every 1 to 2 years rather than every 2 to 3 months.

7.2.3   Cationic-Substance Coated Media

Different types of solid media such as granular activated carbon, clays, and zeolite can be coated with cationic substances and then remove perchlorate from water in an ion exchange process. Research into the use of such "tailored GAC" and other cationic-substance coated media has recently intensified due to the current high cost of regenerating or disposing of ion exchange resins. If the medium coated by the ionic substance can be regenerated inexpensively, then this technology may be more cost effective than ion exchange resins because it would treat almost as many bed volumes of water, but the production costs would be lower because the substrate (granular activated carbon or clay for example) is less expensive to produce. Since ion exchange resins dominate the market for treatment systems that remove low concentrations of perchlorate, these potentially cost-competitive media would have a ready market for their use.

7.2.3.1   Background and Theory

Various solid media either have a high surface area or strong ionic bonding potential. Examples of these include manufactured GAC, with a significant surface area but weak ionic bonding potential [23]; bentonite clay, a form of montmorillonite, with strong ionic bonding potential (cation exchange capacity of 70-90 milliequivalents per gram [meq/g]) [24]; and naturally occurring and manufactured mineral zeolites, which have both significant surface area and moderate bonding potential (2 to 6 meq/g) [25].

Research into cationic coatings for perchlorate treatment media focuses on the family of quaternary amines. Quaternary amines are simply long, folded carbon chains with positively charged ammonium ($-CH_2N(CH_3)_3^+Cl^-$) or similar functional groups at the head of the chain. These substances belong to the larger family of surfactants, as they have both a hydrophilic head and one or more hydrophobic tails. Quaternary amines have a strong affinity for perchlorate but are quite miscible with water under normal temperature and pressure, and so must be bound to a solid medium to remove perchlorate from water.

Research has shown that increasing the number of positive charges on the surface of a solid medium improves the adsorption of perchlorate [22] in the case of GAC.

To do this, manufacturers mix a concentrated solution of the cationic substance with the solid media. The positively charged amine functional groups of the molecules of the cationic substance become physically adsorbed or form ionic bonds with the surface of the solid media.

For example, in the case of bentonite (as well as zeolites), the positively charged amine functional group supplants the sodium, calcium or other positively charged ions at the surface of the solid media. The carbon chain of the quaternary amine molecule then protrudes from the solid medium. The solution of the cationic substance contains more molecules than can stoichiometrically affix to the media surface. Thus, the carbon chains of the cationic substance remaining in solution tend to adhere to the carbon chains fixed to the solid medium by a "tail-tail interaction." This gives the coated bentonite, also referred to as an organoclay, a small positive charge, which allows it to remove some anions, such as perchlorate [24].

The technology has shown promise in the treatment of groundwater containing low levels of perchlorate contamination (less than 100 μg/L) associated with certain military activities, fireworks, and naturally occurring perchlorate [26]. It has also been demonstrated to be effective in treatment of municipal water supply for a moderate range of perchlorate concentrations [27].

7.2.3.2 Application

Pilot studies have not been reported to date for treatment using zeolites or organoclay. However, a field-scale demonstration in 2004 and 2005 at Redlands, California successfully used a polymer-tailored GAC to remove perchlorate from groundwater. The demonstrations showed that tailored GAC reduced perchlorate concentrations from 75 μg/L to less than 4 μg/L using an EBCT of less than 10 minutes. In addition to perchlorate, several anions exist in the Redlands groundwater at high concentrations [18] [22]. These anions can compete with perchlorate for sorption sites on the tailored GAC.

Case Study: Massachusetts Military Reservation, Cape Cod, Massachusetts
    As described previously, groundwater remediation systems at this site extract contaminated groundwater, treat the water using GAC and IX resins, and reinject the water. Laboratory-scale and field-scale studies in 2003 and 2004 using a monomer-tailored GAC rather than a polymer-tailored GAC tested the site-specific performance of tailored GAC on groundwater at MMR [28] [29] [30]. The groundwater used in these studies contained very low concentrations of perchlorate (i.e., less than 10 μg/L).
    The laboratory studies consisted of RSSCTs as described in Section 7.2.2.2. The critical information from the studies is tabulated in the following chart.
    A subsequent field scale study treated groundwater at an average rate of 3 gpm, with an average perchlorate concentration of 3 μg/L in the influent. The treatment unit comprised a standard 75-pound GAC vessel containing tailored GAC, with an EBCT of 5 minutes. The system processed a total of 900,000 gallons of water

Summary of MMR Laboratory Studies of Tailored GAC [AMEC 2004a, AMEC 2004b, AMEC, 2004c]			
Influent Perchlorate Concentration (µg/L)	Type of GAC	EBCT (min)	Bed Volumes to Perchlorate Detection[1]
1	Standard	20	30,000
1	Standard	9	46,000
5.6	Standard	20	20,000-31,000
5.6	Standard	7.5	25,000
5.6	Standard	5	15,000
1	Tailored	9	270,000
5.6	Tailored	5	170,000

[1] Detection limit for perchlorate was 0.35 µg/L

for a total of 60,000 bed volumes. Perchlorate was never detected in the effluent; that is, breakthrough did not occur [30]. As described previously, standard GAC and ion exchange resins were selected for the full-scale treatment system. This is because the tailored carbon has not, to date, obtained State approval for use in groundwater treatment systems.

Tailored GAC may be a cost-effective alternative to perchlorate selective IX resins. The challenge for tailored GAC is the regeneration of the medium. As of early 2006, no large-scale facilities can regenerate tailored GAC. Therefore, any spent tailored GAC must be removed for disposal.

Studies of perchlorate removal using organoclay and zeolites that have been tailored using surfactants such as the quaternary amines are currently in the laboratory stage. No field testing has been performed as of the time this book was written.

7.2.4 Membrane Filtration Technologies

Membrane filtration treatment processes include reverse osmosis (RO), nanofiltration (NF), ultrafiltration (UF), and electrodialysis (ED). Membrane technologies have been shown to be effective in the removal of perchlorate from groundwater, concentrating the perchlorate into a brine solution. Fouling remains a concern in full-scale systems and to date effective technologies for post processing/disposal of brine solutions produced by membrane technologies have yet to be identified. Brine quantities can be as high as 15 to 20% of the volume of groundwater treated [12].

7.2.4.1 Background and Theory

Membrane filtration units force water through a semi-porous polymer membrane to remove dissolved ions such as perchlorate. These contaminants cannot pass through the membrane, and thus concentrate in a waste solution. Membrane filtration processes differ in terms of the pore size of the membrane and the applied pressure. Each of the four general processes (reverse osmosis, nanofiltration, ultrafiltration, and electrodialysis) is described briefly below.

RO treatment units force water through a semi-permeable membrane under high pressure to remove dissolved solids from the influent water. The RO membranes use simple diffusion to reject contaminants such as dissolved ions. High-pressure (greater than 150 pounds per square inch or psi) RO membranes can remove about 99.9% of perchlorate [31]. The perchlorate is concentrated in the RO unit's brine stream. The process produces permeate (that is, clean water) and reject water (in the form of a brine stream) that contains contaminants requiring disposal or additional treatment. With respect to perchlorate treatment, research into the most effective means to process/dispose of the brine stream continues.

In the NF process, water passes through a partially permeable membrane, which rejects perchlorate based on a combination of solution/diffusion, electrostatic charge, and size exclusion [33]. NF membranes have a nominal pore size of approximately 0.001 microns and require typical operating pressures between 80 psi and 150 psi. In contrast to RO membranes, NF membranes allow some particles such as some salts to pass through, and these membranes operate at a lower operating pressure than RO membranes [31].

UF incorporates a membrane with a larger pore size than NF, approximately 0.002 to 0.1 microns. It therefore operates at a lower pressure (30 to 100 psi) and retains contaminants less selectively. Most inorganic ions can pass through a typical UF membrane [32]. "Tight" UF (pores are smaller than in typical UF membranes but larger than NF membranes) has been tested to treat perchlorate but has been shown to be ineffective [33].

In the ED process, water passes through semi-permeable and permeable membranes to separate cations and anions. An electrical field promotes the separation of the cations and anions through the membranes. Testing of the electrodialysis process has shown that low concentrations of perchlorate can be removed in water with high total dissolved solids (TDS) [34].

7.2.4.2 Application

Three main types of semi-permeable membranes have been evaluated for the treatment of perchlorate in groundwater: 1) high-pressure RO membranes; 2) nanofiltration membranes; and 3) low pressure RO membranes [31]. ED has also been tested for perchlorate treatment. However, performance data for membrane treatment are limited. The American Water Works Association Research Foundation (AWWARF) is supporting an ongoing research project to investigate the feasibility of membrane filtration technology for perchlorate removal from water sources of different quality.

Morss [31] evaluated the performance of point-of-use RO systems to treat perchlorate. These systems typically operated under low-pressure to treat tap water. While the data revealed that these systems achieved 95% rejection of perchlorate, the RO unit would likely require an additional polishing step for removal of residual perchlorate in order to meet NSF International[1] standards for potable use.

The Metropolitan Water District of Southern California reported that RO and nanofiltration membranes were capable of removing 80% or more of the perchlorate from 18 μg/L to less than 4 μg/L [35]. Nanofiltration membranes generally require pH shifts at the membrane surface to effectively remove perchlorate [31]. However, fouling of the membrane filters remains a concern for both reverse osmosis and nanofiltration technologies.

An electrodialysis reversal (EDR) pilot unit (7 gpm) was installed at the Magna Water Company in Utah. In EDR, the same membranes used in the ED process are used to provide a continuous self-cleaning electrodialysis process, which uses periodic reversal of the polarity to allow systems to run at higher recovery rates. Fouling and scaling constituents are removed during reversal, sending fresh product water through compartments previously filled with concentrated waste streams. Therefore, EDR systems operate with higher concentrations in the brine or concentrate streams with less flow to waste.

The membranes in the EDR stack at the Magna Water Company were constructed from ion-exchange resin material. The results of the pilot test indicated that, for the water quality observed at Magna, EDR provides effective removal of perchlorate (concentration of perchlorate ranged from 15 to 130 μg/L). Some sorption of perchlorate to the EDR ion-exchange membranes appeared to occur. Removal efficiencies ranged from 70% for a two-stage system to 94% for a four-stage system [34].

The effectiveness of membrane technology is limited to sites where concentrations of perchlorate are low because competing technologies such as ion exchange resin removal or biological destruction are less expensive. Membrane filtration has high operation costs compared to other perchlorate treatment technologies due to membrane fouling caused by biological growth and precipitation of ions. The membrane filtration process also requires that the concentrated brine stream be disposed of or treated. Treatment of the brine stream via biological reduction may be difficult due to the high concentration of salts that may inhibit biological activity. At this time this pre-concentration step is not cost-effective nor does it appear to offer significant benefits over direct treatment with ion exchange or biological reduction.

7.2.5   Capacitive Deionization Using Carbon Aerogel

Capacitive Deionization (CDI) was developed at Lawrence Livermore National Laboratory in 1994 and 1995 [36]. In this removal technology, aqueous solutions containing anions such as perchlorate and cations pass through a stack of carbon aerogel electrodes. (Gels are semisolid materials, composed of matter in a colloidal

---

[1] NSF International is a non-profit non-governmental organization that certifies treatment media for public water supply systems.

state that does not dissolve. Carbon aerogel is an air-filled carbon gel, and is a solid-state substance similar to gels, but where the liquid component is replaced with gas.) The stack of electrodes behaves as a capacitor. Each electrode has a very high specific surface area (400 to 1,100 m^2/g). Water containing cations and anions - in this case, perchlorate - passes through pairs of electrodes, which act as cell channels for the water. The current through the electrodes polarizes the solution: anions such as perchlorate concentrate near the cathode, and cations concentrate near the anode. After polarization, non-reducible and non-oxidizable ions are removed by reversing the polarity of the cell, causing the capacitor to release the contaminants into the cell channels [37]. The contaminants are removed from the cell by flushing with a small quantity of liquid forming a concentrated solution. The effluent from the cell is puri-fied water.

This technology has found little application to perchlorate treatment to date. First, the weak electrochemical force inherent in the technology limits the through-put. Second, the sorption capacity of the carbon-aerogel anodes decreases with the size of the ion. In the case of perchlorate, a relatively large monovalent anion, the electrosorption capacity is less than the capacity for chloride [38]. Finally, the current high cost of carbon aerogel makes it impractical for perchlorate treatment, although the process is used for desalinization.

7.3 Destruction Technologies

Certain treatment technologies can destroy the perchlorate ion. While perchlorate is a strong oxidant, kinetic limitations render perchlorate stable under ambient conditions. Bacteria can reduce perchlorate in soil or groundwater under certain anaerobic conditions. Perchlorate compounds can thermally decompose at temper-atures in the range from 480°F and 1,200°F, or 250°C to 650°C (see Chapter 2), although for practical purposes, destruction does not occur until at least 600°F (315°C).

7.3.1   Bioreactors

A bioreactor is an *ex situ* biological treatment system that degrades contami-nants in extracted groundwater using microorganisms. In order to survive, the microorganisms require a supply of nutrients and an electron acceptor. In general, biological treatment can be aerobic, anaerobic, or a sequencing of the two to treat contaminants and their degradation products. Anaerobic systems are used to treat perchlorate.

7.3.1.1   Background and Theory

Aerobic treatment requires atmospheric or dissolved oxygen, which serves as the electron acceptor in a reduction/oxidation (redox) reaction. Contaminants are oxidized to form carbon dioxide, ammonia, microbial cell tissue, and other byproducts. Because perchlorate is already a highly oxidized compound, aerobic treatment is not effective in destroying perchlorate.

Anaerobic treatment does not require atmospheric or dissolved oxygen. The microorganisms most often used in perchlorate treatment are facultative anaerobes that can use electron acceptors other than oxygen, such as nitrate, sulfate, and perchlorate to reduce organic compounds. The microorganisms differ from aerobic bacteria in that they can use oxygen for metabolic energy from sources other than dissolved oxygen. In water treatment they utilize oxygen based upon the following preferred sequence of electron acceptors [39]:

- Dissolved oxygen
- Nitrate (product is gaseous nitrogen)
- Perchlorate (product is chloride)
- Sulfate (product is hydrogen sulfide)

Biological treatment processes that use facultative anaerobes usually use heterotrophs instead of autotrophs. In addition to the electron sink (oxygen, nitrate, perchlorate, etc.) heterotrophs require an organic carbon source for growth. Typical carbon sources are methanol, ethanol, or acetate. In contrast, autotrophs can take carbon from CO_2 that is present in the water and form mono- and di-saccharides for growth [40].

Studies have shown that most groundwater naturally contains at least some perchlorate reducing bacteria (PCRB) [41]. Under reducing conditions found in perchlorate bioreactors, the percentage of PCRB species will increase over time as the microbial population self-selects for perchlorate as the dominant electron acceptor. The higher the concentration of perchlorate and nitrate in the groundwater, the more quickly the PCRB will become the dominant species until equilibrium is reached for the concentration of perchlorate in the influent. For further discussion of microbial degradation of perchlorate in the presence of nitrate see Section 4.4.

Biological treatment processes have received a lot of research attention for perchlorate treatment. To create anaerobic conditions in the water, operators add a readily biodegradable substrate (such as acetate, ethanol, brewer's yeast extract, molasses, or compost extract) and isolate the water from the atmosphere to prevent oxygenation. Addition of the substrate typically induces the indigenous microbes present in the groundwater to extract electrons from the substrate and deliver them to the perchlorate. This electron transfer induces biotransformation of the perchlorate ion to chlorate and then chloride.

7.3.1.2 Application

Suspended-growth or attached growth bioreactor systems may be used to treat extracted groundwater. In suspended-growth systems, the microorganisms grow in solution and are suspended in the water being treated. The most common suspended growth system, activated sludge, is the classic form of aerobic biological wastewater treatment: most municipal sewage treatment plants utilize activated sludge. The term "activated sludge" refers to the biological mass (sludge) maintained and suspended in a continuous flow stirred tank reactor. (In aerobic treatment, air is sparged into the bottom of the tank to provide oxygen and mixing.) The effluent from the treatment tank flows to a clarifier, where biological sludge settles by gravity. Some of the biological sludge is recycled to the continuous flow stirred tank reactor. The remainder of the sludge is wasted from the clarifier to remove organic material from the system and sustain the operation of the system. The wasted sludge must be dewatered and disposed of appropriately.

In attached growth systems, as the name implies, a film of microbial growth is attached (or fixed) to an inert or active support medium. As the water flows over the support medium, the contaminants are biodegraded. Designs for systems to treat perchlorate include biological fluidized bed reactors (FBRs), packed or fixed bed reactors (PBRs), and hollow-fiber membrane biofilm reactor (MBfR).

Bioreactors are traditionally designed to optimize natural biodegradation using equations based on the food-to-microorganism ratio, reaction kinetics, and/or hydraulic loading and retention time. The design of a treatment system ultimately requires site-specific treatability testing at bench and/or pilot scale. Design considerations include:

- The type of contaminants in the groundwater, concentrations, and potential variations in concentrations over time
- Potential toxicity of groundwater constituents such as heavy metals to the microorganisms
- Influent concentrations of dissolved oxygen and nitrate concentration, which increase the required hydraulic residence time because they are reduced prior to perchlorate being used as an electron acceptor
- Air supply for aerobic treatment or prevention of air infiltration for anaerobic treatment
- The need for supplemental food, or a co-metabolite
- Requirements for macro- and micro-nutrients, including nitrogen, phosphorus, and other compounds
- Requirements for controlled environmental conditions such as adequate temperature and pH conditions to promote biological activity
- Start-up time to generate adequate biofilm
- Formation of harmful by-products during perchlorate degradation
- Potential for air emissions of volatile organic compounds
- Channeling problems in PBRs may require relatively frequent backwashing
- Sludge generation from the dying biomass, dewatering and disposal

The U.S. Air Force (USAF) Research Laboratory, Materials Manufacturing Directorate developed the initial biological application to treat high-level (i.e., 1,000 to 10,000,000 μg/L) perchlorate concentrations in wastewater from a rocket manufacturing facility in Utah [42]. When treatability tests demonstrated the success of the process, the USAF designed and built a production-scale, continuously stirred bioreactor system, which began operating at the facility in 1997.

Adapting the approach to low-level perchlorate concentrations in groundwater, pilot-scale tests at the Aerojet Superfund site in Rancho Cordova, CA in 1996 utilized an anaerobic fluidized bed bioreactor. Those tests demonstrated that the bioreactor system could reduce perchlorate concentrations in groundwater from over 5,000,000 µg/L to the thousands of µg/L [12] [43]. The full-scale system put into operation in 1998 reliably treats the waters to less than 4 µg/L (the detection limit) of perchlorate [12]. Table 7.4 provides additional examples of successful pilot and full-scale anaerobic *ex situ* biological systems.

Table 7.4 Perchlorate Treatment Systems Using Biological Systems
[12] [42] [44] [45]

Site	Perchlorate Concentration (µg/L)	Implementation Stage	Type of Biological System (manufacturer)
Aerojet, Rancho Cordova, CA	8,000	Full Scale	Fluidized Bed Reactor (Shaw Environmental, Inc./US Filter - Envirex)
Thiokol, Brigham City, UT	4,000,000 to 6,000,000	Full Scale	Continuously Stirred Tank Reactor (Applied Research Associates, Inc.)
Longhorn Army Ammunition Plant, Karnack, TX	35,000	Full Scale	Fluidized Bed Reactor (Shaw Environmental, Inc.)
Kerr-McGee, Henderson, NV	350,000	Full Scale	Fluidized Bed Reactor (Shaw Environmental, Inc.)
Hodgdon Powder Co., KS	300,000	Full Scale	Continuously Stirred Tank Reactor (Applied Research Associates, Inc.)
McGregor Naval Weapons facility, McGregor, TX	500-5,000	Full Scale	Fluidized Bed Reactor (Shaw Environmental, Inc.)
DOD Facility, CA	250-400	Pilot Scale	Hall Bioreactor (EcoMat, Inc.)
La Puente, CA (Municipal water supply)	55-1,000	Pilot Scale	Hollow-Fiber Membrane Biofilm Reactor (Northwestern University, Applied Process Technology)
Redlands, CA (Municipal water supply)	70	Pilot Scale	Packed Bed Reactor (Pennsylvania State University)

7.3.1.3 Biological Fluidized Bed Reactor

Biological FBRs, a type of attached-growth system, have been successfully applied in aerobic, anoxic, and anaerobic environments. The biological FBR system consists of a reactor vessel containing a granular medium (typically sand and or activated carbon) that is colonized by active bacterial biomass. The upward flow of wastewater or ground-water into the vessel fluidizes the medium. The medium provides a support to which bacteria attach and grow. Figure 7.3 provides a schematic of a typical biological FBR.

Biological FBR systems typically include the following features:

- An influent stream of contaminated groundwater.
- A granular medium (typically sand or GAC) that is colonized by active bacterial biomass.
- Controlled addition of a nutrient substrate, such as acetic acid (vinegar), denatured alcohol (ethanol), or molasses to provide an electron donor.
- Controlled addition of growth nutrients (nitrogen, phosphorous) and pH control chemicals such as sulfuric acids and sodium hydroxide.
- Hydraulic control to maintain fluidization of the system, by suspending the GAC, and provide enough hydraulic residence time to treat the influent water to desired performance goals.

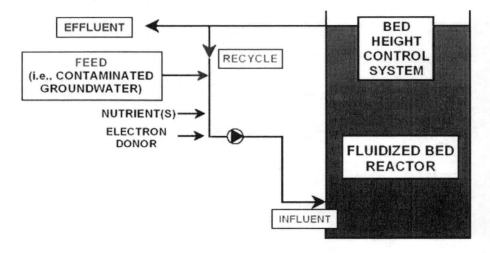

Figure 7.3 Typical Biological Fluidized Bed Reactor System
(Courtesy Shaw Environmental and Infrastructure, Inc.)

- Treated water exiting the reactor, which is recycled or discharged. Because the effluent usually contains low concentrations of biomass it may undergo sand filtration before discharge.

Anaerobic biological FBRs have successfully degraded perchlorate in full-scale applications. A bench-scale treatability study performed at MMR [46] examined whether a biological FBR could degrade both perchlorate and explosives. The test showed that the bacteria in an anaerobic FBR with a nutrient substrate of acetic acid along with phosphorus and nitrogen as nutrients, could successfully degrade perchlorate from approximately 90 μg/L to less than 1 μg/L, with a hydraulic residence time of 35 minutes. The results indicated that this technology most effectively treated groundwater containing perchlorate and explosives above 100 μg/L [46]. Below this concentration, especially when nitrate concentrations in the groundwater are also low, electron acceptors must be added to the system, usually in the form of sodium nitrate, to maintain sufficient biological activity to destroy the perchlorate. This added measure increases costs, which make ion exchange systems more competitive.

Biological FBR systems have been successfully operated in other locations. An anaerobic biological FBR system has operated at the LHAAP in Karnack, Texas since 2001. This system treats influent water containing up to 22,000 μg/L perchlorate to a concentration below the reporting limit of 4 μg/L. The FBR system at GenCorp Aerojet Facility in Sacramento, California currently treats perchlorate in the influent water from 4,000 to 8,000 μg/L to an effluent concentration below the reporting limit of 4 μg/L. A third FBR system, at the Jet Propulsion Laboratory in Pasadena, California, degrades perchlorate from concentrations of 300 to 800 μg/L to below the reporting limit of 4 μg/L [42]. A fourth FBR system has been installed at the Kerr-McGee facility in Henderson, Nevada as described in the accompanying case study.

Case Study: Kerr-McGee and PEPCON, Henderson, Nevada
Fluidized bed reactors (FBRs) gained ascendancy as a means to treat perchlorate starting in 1998, when four units were installed at the Aerojet site at Rancho Cordova California to treat 4,000 gpm, with an increase to 5,300 gpm in 2003. In 2001, as previously noted, a smaller unit was installed at the LHAAP. At the Kerr-McGee site, FBRs were not the first choice of treatment system. Instead, ion exchange resin treatment systems were first selected for use at the site.

FBR Treatment System at Kerr-McGee
(Nevada DEP, 2005 [45])

Initial perchlorate control efforts at the Kerr-McGee site began in 1999, when the company began temporary remedial measures to meet the requirements of the company's Consent Agreement with the Nevada Department of Environmental Protection (NDEP). Where a groundwater seep discharged into the Las Vegas Wash, workers devised a capture system and pumped groundwater to two ion exchange units. The influent to the ion exchange units comprised approximately 300 gpm and contained perchlorate concentrations of 80,000 to 100,000 μg/L. Based on the available technology, treatment entailed a once-through ion exchange process. A four-well system was installed in 2001 and was increased to nine wells in 2003, with a corresponding treatment flow of 600 gpm. This treatment removed over 90% of the initial perchlorate, to concentrations between 500 and 2,000 μg/L. In 2003, Kerr-McGee replaced the single pass system with an ISEP™ system, with an increased capacity of 850 gpm to treat perchlorate influent concentrations of up to 300,000 μg/L. The treated water still contained concentrations of perchlorate of 500 to 2,000 μg/L [12].

By 2003, the effectiveness of biological treatment of perchlorate was well known. It was also becoming clear that ion exchange resins could effectively remove perchlorate from groundwater to concentrations on the order about 1 μg/L when the initial concentrations of perchlorate are about 10,000 μg/L, but not higher. The biological FBR system had already proven itself at LHAAP, and was therefore chosen for piloting at the Kerr-McGee facility along with a continuously stirred treatment bioreactor (CSTR). Both technologies successfully destroyed perchlorate, reducing perchlorate from influent concentrations of 350,000 to 400,000 μg/L to less than 4 μg/L. However, construction of the CSTR system would cost significantly more than the FBR system. Kerr-McGee therefore selected the biological FBR for full-scale implementation. Installed in 2004, the system now treats influent flows of approximately 1,000 gpm to less than the NDEP's provisional action level of 18 μg/L [12].

7.3.1.4 Biological Fixed Bed Reactor

Packed/fixed bed reactors or PBRs are conceptually similar to FBRs except that they rely on the downward flow of water through the treatment column, as seen in trickling filters used for wastewater treatment. PBRs support the biomass on stationary sand or plastic media with a large surface area for bacterial attachment. The contaminants degrade when they diffuse into the biomass. In some cases PBRs use hydrogen gas to promote reducing conditions to support anaerobic microorganism growth. The hydrogen gas-based systems, which were developed at the bench scale but have not been implemented at full scale, promote chlorinated aliphatic reduction by hydrogen-oxidizing bacteria under highly reducing conditions [47].

PBR systems offer some advantages over FBRs. In general, they are simpler to operate and do not require the same level of power as FBRs to keep the system fluidized [42]. However, they can also have disadvantages. In general, they are not as efficient as FBRs, allow less control over the treatment process, and can be subject to icing in cold weather unless enclosed. In addition, FBRs have high surface area for biomass attachment and growth, and low pressure drop across the bed [12]. Further, biomass buildup on the support media requires that the packed bed system be periodically backwashed to prevent clogging [48]. Continuously self-cleaning fixed-film (CSCF) bioreactors, developed to address food and beverage processing wastewater as well as wastewater from corrugated cardboard manufacturing lines, are in full-scale operation in Japan [49]. Adaptations to the technology for additional applications could resolve the problems associated with the biosolids buildup and plugging on packed bed reactors [50].

Treatability tests of anaerobic PBRs have been conducted on moderate- to low-level (< 1,000 μg/L) perchlorate concentrations in groundwater by engineering companies; government entities such as NASA and the Naval Facilities Engineering Center; and by researchers at Pennsylvania State University and the University of Nevada. In each case, the PBR was amended with acetate and/or nutrients and/or inoculated with *Dechlorosoma sp. KJ* [39] [47] [51] [52]. Promising results were reported. These results could signify that a full-scale PBR system could be capable of treating perchlorate-impacted waters.

Field tests in 2002 at the Texas Street Well Facility in Redlands, California utilized sand filters and plastic media, with flow rates of 1 to 2 gpm, to treat perchlorate contaminated groundwater. The systems destroyed influent concentrations of perchlorate of 77 μg/L to levels below a detection limit of 4 μg/L. Complete biological destruction of perchlorate required a detention time of 10 to 18 minutes in sand filters, but 60 minutes in the reactor containing plastic media.

A fixed bed system currently under development using autotrophic microorganisms uses an up-flow bed of sulfur granules. The resulting system is not fluidized; the bed therefore remains intact. The microorganisms in the system self-select for the environment created by the sulfur, using the sulfur as an electron donor (with CO_2 as the carbon source, as noted previously) and converting it to sulfate in the process of reducing the perchlorate ion. Originally developed at Kiwa, the national research center for the Netherlands, this process has been demonstrated to be effective in the removal of nitrate in a number of locations [53]. Tests on perchlorate removal using this technique have begun on the laboratory scale at the University of Massachusetts and the University of Notre Dame [54] [55].

7.3.1.5 Hollow Fiber Membrane BioFilm Reactor

Researchers at Northwestern University developed a hollow-fiber membrane biofilm reactor (MBfR) to treat perchlorate. The technology uses molecular hydrogen as an electron-donor substrate instead of more complex donors such as acetic acid or ethanol.

In a MBfR, influent water is pumped upwards through a column. Hydrogen is separately pumped into hollow fiber membranes placed vertically within the column. Those membranes provide a solid surface to support bacterial growth. The hydrogen fills the inside of the fibers and passively diffuses through the membrane. The passive diffusion process keeps the hydrogen dissolved in the water, making it more available to the biofilm growing on the outside of the hollow fibers. The biofilm usually develops from indigenous bacteria present in the ground-water [56] [57]. Figure 7.4 illustrates the bioreactor configuration.

Advantages of the MBfR include less production of biomass than the FBR and PBRs and stoichiometric utilization of the electron donor in the reduction process, which means that less hydrogen is required than other electron donors.

Pilot tests demonstrated the MBfR technology for nitrate and perchlorate removal at La Puente California as part of an AWWARF project in 2002 [58]. During the initial laboratory bench-scale study, treatment in the MBfR reduced perchlorate concentrations from 1,000 μg/L to the detection limit of 4 μg/L. In a follow-on pilot scale study at La Puente, the perchlorate concentrations were reduced from 55 μg/L to below the detection limit of 4 μg/L, with a hydraulic residence time between 15 and 60 minutes. Treatment also reduced nitrate concentrations.

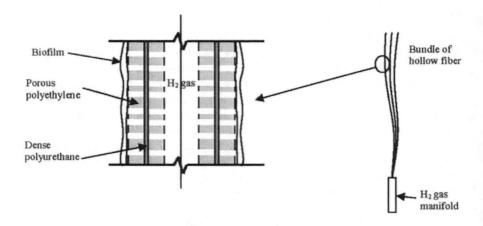

Figure 7.4 Hollow Fiber Membrane BioFilm Reactor
(Courtesy Applied Process Technology, Inc.)

7.3.1.6 Biofilter Denitrificaton

Since the development of the fluidized bed bioreactor, other bioreactors have been developed and tested in the field. One design provides a fluidized system simi-lar to the biological FBR described previously, with two major differences.

First, the system includes a pre-treatment de-aeration step using bio-balls. Bio-balls are small, plastic balls with a high surface area to volume ratio to support the biological media in wet/dry filters. The biofilm growing on the bio balls reduces the dissolved oxygen concentration to 0.5-1.0 mg/L, the optimum concentration for either denitrification or perchlorate remediation [59].

Second, instead of using sand or GAC for the substrate for microbial growth, the bioreactor (patented under the title of a Hall Bioreactor) uses polyurethane-based sponges cut into one-centimeter cubes. The sponges, like the bio balls,

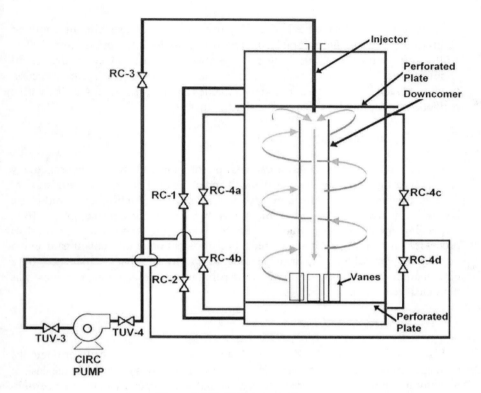

Figure 7.5 Hall Bioreactor Figure
(Courtesy P.J. Hall, EcoMat, Inc.)

have a high surface area to volume ratio. They last for up to several years, and are kept capable of supporting microorganisms by virtue of their gentle collisions with each other and with the walls of the reactor, sloughing off dead bacteria with each collision. Figures 7.5 and 7.6 depict this bioreactor.

Figure 7.6 EcoLink Polyurethane Sponges
(Courtesy P.J. Hall, EcoMat, Inc.)

A pilot system was tested at a Department of Defense site in southern California in 2000. Perchlorate concentrations at the site varied from 300 to 1,000 μg/L, with the month-long pilot test average influent concentration of about 350 μg/L. Methanol was used as the carbon source. The system successfully destroyed perchlorate to non-detectable levels (i.e., less than 4 μg/L) with a residence time of approximately one-half hour [59].

7.3.2   *In Situ* Biodegradation (ISB)

In situ biodegradation (ISB) combines microbiology, chemistry, hydrogeology and engineering into a strategy for planned and controlled microbial degradation of specific contaminants. ISB can be deployed for source reduction, dissolved-phase contaminant reduction, or as a biological barrier to contain the plume [60]. Depending on the contaminants(s), site conditions, and remediation goals, ISB processes can be designed based on reduction or oxidation of the contaminant, either directly or cometabolically, or, depending on the transformation sequence, by a combination of reactions. In the case of perchlorate, the degradation pathway is biological reduction.

7.3.2.1 *Background and Theory*

ISB typically involves the delivery of nutrients to the subsurface to promote the biodegradation of the target contaminants by the indigenous bacteria. In the case of perchlorate, which degrades reductively, the nutrients typically consist of carbon-based electron donors such as alcohols (ethanol), organic acids (lactate, acetate, citrate), sugars, or edible oils (canola). As in the *ex situ* bioreactors described previously, *in situ* bacteria use the electron donors as a food source, while using the perchlorate as an electron acceptor. In this process, microbes break down the contaminants to benign end products such as carbon dioxide (CO_2), water, and chloride.

To provide a basis for discussing the *in situ* treatment of perchlorate, this section briefly describes techniques for *in situ* bioremediation. It is adapted from *Fundamentals of Hazardous Waste Site Remediation* [1], with permission.

Treating groundwater contaminants in place can offer many advantages over pump-and-treat systems. In a pump-and-treat system, potentially costly equipment and energy are needed to bring groundwater to the surface and move the water through a series of treatment units. The options for discharging treated water are sometimes quite limited. However, *in situ* treatment also has limitations. The mass transfer of reagents and contaminants can be limited in heterogeneous or low-permeability soil. *In situ* processes can be difficult to control and to monitor. There are several forms of *in situ* treatment, including natural attenuation, biodegradation, air sparging, reductive dehalogenation, oxidation, and surfactants, with the most effective one for perchlorate destruction being biodegradation.

Subsurface conditions must be changed to actively treat groundwater *in situ* using bioremediation. Techniques that may be used for perchlorate treatment include injection wells, recirculation well systems, and permeable reactive barrier systems. Aqueous solutions of nutrients that enhance perchlorate biodegradation can be added to an aquifer through injection wells. Single-well systems can inject reagents and circulate groundwater vertically around each well. The groundwater circulation creates a miniature reactor around each well, providing control of mixing and hydraulic retention time. Alternatively a vertical permeable barrier can be placed perpendicular to groundwater flow to a width and depth sufficient to intercept the entire plume of contaminated groundwater. The contaminated groundwater flows through the barrier of reactive material, which removes or destroys contaminants such as perchlorate.

An aquifer is a complex ecosystem. It contains a variety of microorganisms which compete for food and strive to reproduce. Their environment reflects a complex mixture of variables: contaminant (food) distribution; pH; the type of soil and resulting nutrient levels, hydraulic conductivity, and geochemistry; and the availability of electron acceptors. These variables affect the occurrence and rate of contaminant biodegradation. In order to grow and reproduce, microorganisms need food to supply energy and carbon for cell growth. As described previously, the microorganisms that destroy perchlorate require an anoxic environment, and significant reducing conditions.

Two of the most critical components of the aquifer ecosystem are dissolved oxygen and pH. Temperature is also an important characteristic, with biological activity being very slow at temperatures lower than 10°C, but most groundwater temperatures are above this threshold. Groundwater naturally contains some dissolved oxygen as a result of the contact between air in the atmosphere and precipitation before it infiltrates. The level of dissolved oxygen depends in part on the type of soil through which precipitation must infiltrate. The concentration of dissolved oxygen in shallow groundwater is typically relatively low in silty or clayey soils (i.e., less than 0.1 mg/L) and somewhat higher (i.e., greater than 0.1 mg/L) in sandy or gravelly soils. The concentration of dissolved oxygen also depends on the extent of geochemical reactions in the aquifer which consume oxygen. Finally, the concentration of dissolved oxygen depends strongly on the concentration of organic compounds and the resulting extent of aerobic biodegradation.

In aerobic aquifers with high concentrations of dissolved oxygen, the aquifer must be manipulated to create the reducing conditions required for perchlorate destruction. When the supply of dissolved oxygen becomes depleted, microorganisms begin to use nitrate (NO_3^-) as the electron acceptor. After this point, perchlorate becomes the electron acceptor of choice. When those electron acceptors are used up, microorganisms turn to sulfate ($SO_4^=$) as the electron acceptor, producing hydrogen sulfide (H_2S). Finally, when no other electron acceptors are available, methanogenic organisms use carbon dioxide or certain organic compounds as the electron acceptor and produce methane.

Biological activity is possible at pH levels between 5.0 and 9.0, and optimal conditions are between 6.0 and 8.0. If the pH is too low, a condition that is common in the northeastern U.S., a dilute solution of a base such as sodium hydroxide can be injected into the aquifer before the nutrients are added. In the less common situations where the pH is too high, sometimes no addition may be required, as some of the common nutrients, such as acetic acid, can be added to the aquifer and will provide not only the nutrient basis for biological growth but also lower the pH of the aquifer to levels needed for biological activity. In other cases, acids such as concentrated hydrochloric acid or sulfuric acid may be injected into the aquifer.

Some aquifers contain very low microbial populations. In these cases, perchlorate destruction is only achieved after significant injections of nutrients to encourage microbial growth or with the help of added perchlorate-reducing bacteria. In these cases, the injection well technology, specifically called bioaugmentation, injects not only nutrients but also naturally occurring PCRB [61] [62] to destroy the perchlorate.

7.3.2.2 Application

The team designing an *in situ* treatment system must understand the type of microorganisms, contaminant, and hydrogeological conditions at the site in detail. Since *in situ* conditions are manipulated by engineering means, the most important consideration is the ability to transmit and mix liquids in the subsurface [60]. Successful implementation requires a system of monitoring wells to monitor, control, and verify performance.

Electron donors can be delivered to the aquifer through a variety of active (recirculation) or passive (injection or trenching/barrier) methods. Design of the system is site-specific, depending on site characteristics (plume depth and width, hydraulic conditions) and remedial goals. Major system components include extraction wells, conveyance piping, electron donor dosing station, several electron donor delivery/recharge wells, barrier wall width and depth, barrier components, and performance monitoring wells (used to measure oxidation reduction potential, dissolved oxygen, parent and by-product concentrations, carbon source, microbiological growth). Operation and maintenance include operator labor for refilling electron donor tanks, and performance monitoring. Periodic rehabilitation of electron donor delivery wells may be required as fouling can reduce the permeability of the formation and amendment delivery wells.

In situ perchlorate treatment may provide advantages over *ex situ* treatment, because treating groundwater without pumping it to the surface can save substantial cost. However this advantage is limited to sites where the groundwater is within reasonable depth limits and the aquifer conditions favor perchlorate treatment. If the aquifer is highly aerobic, very low in cometabolites or nutrients, and/or the perchlorate plume is widely dispersed, the expense of injecting sufficient materials to support the growth of PCRB may be very high compared with *ex situ* treatment.

To date, *in situ* bioremediation has been applied to many sites to treat perchlorate in groundwater, as shown in Table 7.5 [12]. These *in situ* bioremediation field/pilot demonstrations have shown that perchlorate can be biodegraded below actions levels of 2 to 4 μg/L.

Table 7.5 Summary of Pilot/Field Scale Results for
In Situ Bioremediation [12] [42]

Site and Location	Scale and Type of System	Perchlorate Concentration	Injected Materials	Impact
NASA Jet Propulsion Lab Pasadena, CA	Pilot - Injection	>500 µg/L	Corn syrup	Corn syrup prevented sulfur byproduct buildup
Area 20 Aerojet, Sacramento, CA	Pilot- Injection	12,000 µg/L	Acetate, lactate	Perchlorate < 4 µg/L; Manganese increased under reducing conditions
Area 20 Aerojet, Sacramento, CA	Pilot - Biobarrier	8,000 µg/L	Ethanol	Perchlorate < 4 µg/L
WNN Area, Aerojet, Sacramento, CA	Pilot - Active Biobarrier	3,000 µg/L	Ethanol	NA
Area 41 Aerojet Sacramento, CA	Anaerobic Composting	23 mg/kg	Cow manure, calcium magnesium acetate	(vadose zone) < 0.1 mg/kg in 7 days
Hogout Facility, Aerojet Sacramento, CA	Pilot - Biobarrier + infiltration flushing	NA	Oleate, calcium magnesium acetate	NA
Naval Weapons Industrial Reserve Plant, McGregor, TX	Full - Trenching	27,000 µg/L (also 500 mg/kg)	Wood chips, vegetable oil, cottonseed	Perchlorate < 4 µg/L
Rocket testing facility, NV	Pilot - Groundwater recirculation	540 µg/L	Citric acid	NA
PEPCON (former AMPAC) facility, NV	Pilot - Groundwater recirculation	600,000 µg/L	Ethanol, citrate	2003 - Perchlorate < 2 µg/L
Longhorn Army Ammunition Plant, Karnack, TX	Pilot - Passive Biobarrier	1,000 µg/L	Lactate	NA
Longhorn Army Ammunition Plant, Karnack, TX	Pilot - Composting	350 mg/kg	Chicken manure, cow manure, ethanol	Non-detection after 10 months

Table 7.5 Summary of Pilot/Field Scale Results for
In Situ Bioremediation, continued [12] [42]

Site and Location	Scale and Type of System	Perchlorate Concentration	Injected Materials	Impact
Los Alamos National Laboratory, Mortandad Canyon, NM	Full Permeable Biobarrier	120 to 350 µg/L	Pecan shells, cottonseed, apatite, limestone	NA
Undisclosed site, Northern Maryland	Pilot - Permeable Biobarrier	10,000,000 µg/L	Emulsified soybean oil	NA
Former road flare manufacturing facility, Santa Clara County, CA	Pilot - Infiltration	NA	Acetate, citric acid	NA
Edwards AFB	Pilot - Biobarrier	NA	Food-grade edible oils	NA

NA = Not available

Case Study: Indian Head Naval Surface Warfare Center, Indian Head, Maryland

Several *in situ* bioremediation field studies have been performed at the Indian Head Division, Naval Surface Warfare Center at Indian Head, Maryland (Indian Head). The high concentrations of perchlorate coupled with the relatively shallow groundwater table make this a potentially ideal site for *in situ* bioremediation.

A laboratory study was performed to determine whether humic substances naturally supported the type of microorganisms or perchlorate-reducing bacteria (PCRB) that could biologically degrade perchlorate [63]. Soils were collected from several sites, including sediment from Indian Head. The results of the study showed that indeed many but not all PCRB can use humic substances as electron donors to successfully degrade perchlorate, and that these PCRB are ubiquitous and varied.

A set of laboratory scale studies was performed as part of the under the auspices of the U.S. DoD's Strategic Environmental and Research Development Program (SERDP, in partnership with the Department of Energy and the U.S. EPA) to determine whether PCRB naturally occur in soil, and in what circumstances *in situ* biodegradation could take place [64]. Soils were collected from several locations, including two from Indian Head. The results of the first bench study on soil from the Building 1190 area showed that PCRB do occur naturally in soil, and these PCRB can use injected electron donors such as acetate or naturally occurring humic substances to successfully degrade perchlorate.

The second bench study, performed on soil from the Building 1419 area used methanol, acetate, benzoate, lactate, sucrose, molasses, and a mixture of ethanol and yeast extract in a soil-groundwater slurry with a perchlorate concentration of 45,000 µg/L. The study was performed at a pH of 4.3, which was the natural pH of the soil. The results showed that the low soil pH severely limited the biological reduction of perchlorate.

A follow-up field demonstration of *in situ* bioremediation at the Building 1419 area used recirculating wells [65] [66]. The pH of the natural groundwater was less than 5.0 in the study area, which the researchers determined was too low to support biological activity. The depth to groundwater was between 6 to 16 feet. The project team injected lactate to stimulate naturally occurring microorganisms. In addition, they injected a neutralizing carbonate buffer to counteract the groundwater acidity. Perchlorate concentrations were reduced from about 170,000 µg/L to less than 10,000 µg/L in eight of the nine test plot monitoring wells over five months' time.

This study was the first major *in situ* field study performed at Indian Head, and others are planned for implementation. A description of the studies is provided in the table below.

Summary of Bench and Pilot Scale Testing of Anaerobic Biodegradation, Indian Head Division, Naval Surface Warfare Center

Medium	Scope	Results	Ref.
Sediment	Bench scale testing of one Potomac River sediment sample, using reduced humic substances (HS) as the electron donor, nitrate as the electron acceptor.	Hydrogen sulfide-oxidizing, nitrate-reducing bacteria present in the sediment.	[63]
Groundwater	In situ treatment: groundwater extracted, dosed with lactate (electron donor) and pH buffer, and reinjected. 20 week study.	Microcosm studies showed that low pH (~4) inhibited degradation. Test plot initially contained perchlorate at ~170,000 µg/L (average); levels decreased in most wells by $\geq$ 95% in 5 months (see text for more information)	[65]

Summary of Bench and Pilot Scale Testing of Anaerobic Biodegradation, Indian Head Division, Naval Surface Warfare Center, continued			
Medium	**Scope**	**Results**	**Ref.**
Vadose zone soils	Determine most effective electron donor in column studies. In field pilot test, deliver electron donor to soil via engineered infiltration gallery (shallow soils) and potentially an injection well (deeper soils).	Project began in 2004, slated for completion 2007.	[65]
Vadose zone soils	Determine most effective electron donor in column studies. In field pilot test, deliver electron donor to soil by mixing agents with top 2-3 feet of soil and then watering.	Project began in 2004, slated for completion 2007.	[67]

7.3.3   Phytoremediation

The general term "phytoremediation" describes various *in situ* mechanisms by which vegetation is used to treat hazardous wastes. Some forms of phytoremediation result in the destruction or degradation of the contaminant while others result in the uptake of the contaminant into the plant roots, stems, and/or leaves. In the cases where the contaminant is taken up into the plant system, the plant must be harvested and disposed of or incinerated to remove the contaminant from the site.

Phytoremediation for treatment of perchlorate gained attention in the late 1990s, and was considered for treatment of perchlorate in surface water and groundwater, but no full-scale remediation has been implemented as of early 2006.

7.3.3.1 Background and Theory

The two most common mechanisms for phytoremediation are:

- Rhizosphere degradation - The rhizosphere is the portion of the soil adhering to the root system of a plant. It contains higher microbial numbers, biomass, and activity than surrounding soil, and can therefore provide sufficient microorganisms to biodegrade hazardous organic compounds. However, only contaminants that are biologically available for adsorption to the plant roots or associated microorganisms, including perchlorate, will be degraded using this process.
- Phytoaccumulation or phytoextraction - Heavy metals or inorganics such as

perchlorate are taken up into easily harvestable shoots or trees, which must then be disposed of properly after they are harvested.

As described in Chapter 5, perchlorate can be taken up in plants, especially those that absorb significant amounts of water such as lettuce. Once taken up, the perchlorate accumulates in the leaves and stems as a result of evapotranspiraton, since the water evaporates but the perchlorate does not. In addition, as discussed earlier in this chapter, under anoxic conditions certain microorganisms can degrade perchlorate. Therefore, perchlorate can be degraded or removed from soil or water (using hydroponics) via phytoremediation.

7.3.3.2 Application

U.S. EPA laboratory studies tested dozens of species for perchlorate uptake and degradation [68]. Five factors affected perchlorate uptake: perchlorate concentration, growing conditions (amount of moisture available), nutrient concentrations, plant maturity, and the presence of competing chloride ions. Several species showed promise for further research, including:

- Sweet gum, eastern cottonwood, and black willow for contaminated soils with shallow ground water accessible to plant roots
- Blue-hyssop, smartweed, and perennial glasswort for saturated or inundated wetlands
- Parrot-feather and fragrant white water-lily for water bodies, or for ponds created artificially for phytoremediation
- Tarragon for mechanized flow-through systems where groundwater is extracted, exposed to these plants, then reinjected into the aquifer

Laboratory testing at LHAAP demonstrated that hybrid poplars can remove ammonium perchlorate from a hydroponic solution of simulated groundwater containing perchlorate. In addition, the laboratory studies showed that under reducing conditions facultative bacteria can reduce perchlorate and chlorate to chloride ions and dissolved oxygen, innocuous end-products [69].

Laboratory studies at the University of Georgia for the Wright Patterson Air Force Base [70] showed that rhizosphere degradation can be effective given enough time for the PCRB to develop and multiply. However, high nitrate concentrations appeared to inhibit perchlorate degradation. The most successful plants included French tarragon, cottonwood, and willow. For example, under hydroponic conditions willows degraded perchlorate from 10,000 μg/L to below detection (2 μg/L) in about 20 days, and from 100,000 μg/L to below detection in about 53 days [71]. Although a detectable concentration of perchlorate remained in the leaves upon final harvesting, almost 99% of the perchlorate had been degraded.

Other plants that have been successfully tested include salt cedar trees [72], bulrushes, cattails, and sedges [73]. Phytoremediation using bulrushes, cattails, and sedges was tested on Site 300 at the Lawrence Livermore National Laboratory in

California. There, U.S. Department of Energy (DOE) scientists constructed containerized wetlands systems under low-flow conditions (5 to 10 gallons per minute) in remote locations. Microbial populations in the miniature wetlands degraded both nitrate and perchlorate. Specifically, perchlorate concentrations decreased from 10 to 20 μg/L to less than 4 μg/L.

7.3.4   Thermal Destruction

Perchlorate destruction using thermal methods has been demonstrated for both soil and groundwater treatment. Perchlorate in regenerant brine from the ion exchange process can be thermally decomposed at elevated temperature and pressure with the addition of reducing agents and promoters. Perchlorate concentration of the brine with reverse osmosis or evaporators would be necessary to make the process cost-effective.

Perchlorate in soil can be thermally destroyed in a rotary kiln or other heating unit. Destruction of perchlorate requires temperatures between 600°F and 1,200°F (315 to 650°C) [74] depending on the compound containing the perchlorate. Chapter 2 provides more information regarding temperatures related to perchlorate destruction.

In addition, incineration is a favored process for disposal of ion-exchange resins containing perchlorate. Although the spent resins might be placed in landfills, many generators are concerned with future liability and therefore prefer incineration.

7.3.4.1 Background and Theory

Thermal desorption typically removes contaminants from the soil to the vapor phase and then destroys them. Temperatures between 500 to 1,100°F (260 to 593°C) are used to physically separate moisture and contaminants from the soil. The process is also time dependent in that the contaminants volatilize over a period of time once a target temperature is achieved. The exhaust from this process is collected by an air cleaning system and heated to temperatures of approximately 1,500°F (816°C) to destroy contaminants or otherwise treated.

7.3.4.2 Application

Thermal desorption units commonly incorporate a rotary kiln to mix and heat the soil. Soil is fed into a primary treatment unit, essentially a rotating drum. The soil in the drum is heated to 212°F (100°C) to dry the soil, and then to the target temperature to volatilize contaminants from the soil to be caught in the exhaust or off-gas. The continuous tumbling ensures all of the soil is exposed to the heat for the appropriate time period. Further heating destroys contaminants that are captured in the off-gas while treated soil is discharged from the primary treatment unit.

Treatment of the contaminants continues as the off-gas flows through the air cleaning system, which typically consists of a cyclone, heating unit, and air filter. As the off-gas spins through the cyclone, the soil particles caught in the off-gas are removed before the off-gas moves into a thermal oxidizer. Using temperatures of up to 2,000°F (1,093°C) the heating unit destroys any remaining contaminants, transforming them into nitrogen, water, carbon dioxide, chloride, and other benign products. In a cooling chamber, clean water is used to cool the off-gas to a temperature that can be handled by the filtering system. Air filters then pull any remaining treated particles of soil out of the air. A sonic pulse can be used to periodically disengage the collected particles from the filter and deposit them in the treated soil stream. The filtered off-gas is then discharged.

Case Study: Massachusetts Military Reservation, Cape Cod, Massachusetts

A thermal treatment system was used in 2004 to destroy perchlorate, along with other contaminants, in approximately 66,000 tons of soil from Demo 1 and other areas [74]. Initially, the soil contained perchlorate at concentrations between 1 and 110 μg/kg, with concentrations below 10 μg/kg in about half of the soil volume, and only 10% of the soil volume containing concentrations above 50 μg/kg. Treatment reduced these concentrations to below the performance goal of 4 μg/kg.

Before work could begin on site, the MassDEP required an air emission permit. An application was submitted on September 26, 2003. MassDEP granted "Interim Approval" of the Air Permit on October 30, 2003 authorizing the construction and operation of the Thermal Treatment Unit (TTU) for clean soil only. Subsequently, the MassDEP granted "Conditional Approval" to operate the TTU with contaminated soil on February 20, 2004.

To prepare the site, the project team constructed access roads, a concrete containment pad for the TTU equipment, and a polyethylene-lined asphalt pad to hold the feed soil. They also installed utilities needed for the work: electric power, a water production well, a supply of propane to fuel the thermal process, and a fire suppression system. Then the TTU equipment was mobilized to the site. The TTU consisted of the following components: a soil feed and screening unit, rotary drum dryer, a cyclone dust collector leading to a thermal oxidizer, and an evaporative cooling chamber, a pugmill, and a 20-module bag house.

An initial TTU shakedown and spike testing (i.e., adding a known quantity of perchlorate and testing the removal efficiency) were performed after the equipment was assembled at the site. At the end of the shakedown period, a Proof of Performance test was performed over a three-day period in Spring, 2004. During the test, the TTU system met the project-specified treatment criteria for soil, and met the air emissions criteria established by the Air Permit. Soil treatment criteria were 120 μg/kg for explosives, 2,500 μg/kg for nitroglycerin, and 4.0 μg/kg for perchlorate.

Based on the results from the performance test, the following parameters were established to meet project and Air Permit performance standards:

- Soil treatment temperature: 839°F (448°C), minimum
- Soil feed rate: 40.4 tons per hour (tph), maximum
- Oxidizer discharge temperature: 1,489°F (810°C) minimum

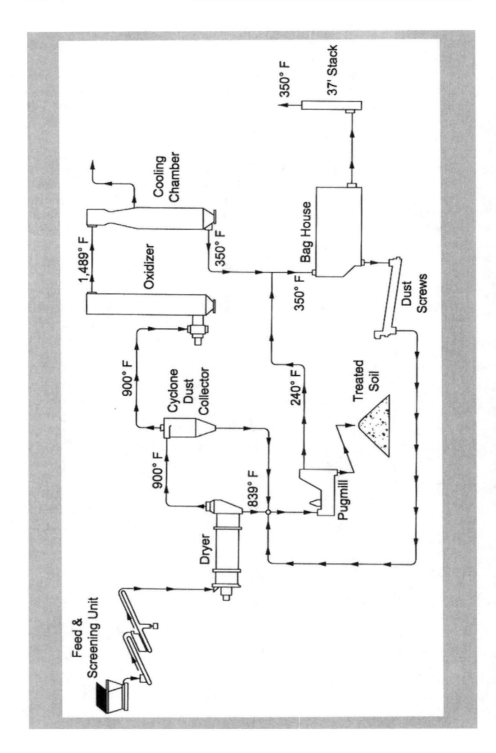

During the laboratory scale treatability study, perchlorate was successfully removed from the soil at temperatures of 775°F (413°C). However, during initial operations, several piles of treated soil failed to meet the 4.0 μg/kg treatment criterion for perchlorate. All soils were successfully treated for explosives. To reduce the re-treatment rate, the treatment temperature was increased to 950°F (510°C). Soils that were not adequately treated in the first pass through the system were retreated.

The process on-line availability was 75% overall for the project (including shakedown) and 80% during full-scale operations. After optimization, the retreatment rate was 1.33%. A total of 65,861 tons of soil were processed during the entire project, which comprised a total operational time for the TTU of 2,160.5 hours. The average processing rate was 30.5 tph for the project and the average treatment rate was 20.6 tph.

7.3.5   Other Destruction Technologies

Several additional technologies have been studied over the past 5 to 10 years, with less success than the technologies cited above. A few of these technologies are briefly described below for the reader's information. More technologies are sure to arise in the next 5 to 10 years as the interest in perchlorate remediation remains high.

7.3.5.1 Composting

Composting is a biological process in which microorganisms convert contaminants such as perchlorate to less harmful byproducts. Heat produced by microorganisms during anaerobic degradation of the contaminants in the waste increases the temperature of the compost pile. Contaminated soil is excavated and mixed with bulking agents such as wood chips or hay, and organic amendments such as manure, and vegetative (e.g., cranberry mash) wastes.   Selection of proper amendment is based on their local availability and their ability to provide a balance of carbon and nitrogen to promote thermophilic, microbial activity. Composting has been performed using three types of process designs: aerated static pile composting (compost is formed into piles and aerated with blowers or vacuum pumps), mechanically agitated in-vessel composting (compost is placed in a reactor vessel where it is mixed and aerated), and windrow composting (compost is placed in long piles known as windrows and periodically mixed with mobile equipment). In addition, composting materials are used to treat groundwater in shallow (less than 25 feet deep) permeable reactive barriers [12].

In 1999, a pilot demonstration of anaerobic composting of soils was performed at an Aerojet facility where perchlorate concentrations were as high as 2,100 mg/kg. The remedial goal was to prevent perchlorate impacts to surface water via overland flow during storm events. Containment consisted of a plastic liner beneath the pile and clean soil berms around the circumference. A plastic tarp covered the top of the five-foot wide by seven-foot high piles to enhance anaerobic, thermophilic conditions. Laboratory treatability studies had shown that using a 1:1 soil to compost ratio, perchlorate was reduced from 700 mg/kg to 0.24 mg/kg in about 90 days, and using a 1:10 soil to compost ratio, perchlorate was reduced from 100 mg/kg to <0.1 mg/kg in about 70 days. Field studies showed that perchlorate concentrations were reduced from an average of 170 mg/kg to less than 0.64 mg/kg in about 38 days [75].

In a similar demonstration at another Aerojet site, a field-scale composting study treated 20 cubic yards of soil. The average perchlorate concentration of 23 mg/kg decreased to about 0.1 mg/kg in seven days. Other studies are being planned or are underway for areas with high concentrations of perchlorate in surface soils.

7.3.5.2 Electrochemical Reduction

This process uses electrochemical reduction, sometimes with a catalyst, to reduce perchlorate to the chloride ion. The process is limited by the available sorption (and thus reactive) sites on the electrode required to capture and reduce the perchlorate ion. Like non-selective ion exchange resins, sulfate and chloride ions are preferentially adsorbed over perchlorate to the reaction sites. In laboratory experiments, this technique reduced only 1% of an initial perchlorate concentration of 4,000,000 μg/L and 35% of an initial perchlorate concentration of 40 μg/L. Currently, this treatment process does not hold great promise as a perchlorate treatment technology [76] [77].

7.3.5.3 UV-Catalyzed Iron Reduction

Metallic iron (zero-valent iron or Fe^0), goethite ($FeO \cdot OH$), or other metals can be used to reduce perchlorate. When the iron dissolves, it releases electrons that are attracted to and can reduce the perchlorate ion. However, the reaction rates are generally quite low due to perchlorate's large activation energy. The reaction rate can be enhanced by changing the pH, temperature and surface area of the iron, but even more so by adding a catalyst to provide sufficient energy. This energy can be in the form of heat or light. Perchlorate absorbs ultraviolet (UV) light in the wavelength range shorter than 185 nanometers, and therefore UV light can be used to catalyze the reduction reaction [78].

In laboratory studies where UV light catalyzed the reaction of perchlorate with Fe^0 in an anoxic environment, [79] perchlorate degraded to chloride and water. The rate of perchlorate reduction depended on the concentration of Fe^0 and the intensity of the UV light. In the experiment, a 1,000 μg/L solution of perchlorate was reduced by 77% in 3 hours, using 100,000 mg/L Fe^0 and in the presence of a UV intensity of 0.9 watts/cm^2.

The limited data on perchlorate reduction with metallic iron and UV light indicate that the technology cannot remove perchlorate to below typical regulatory limits at this stage in its development. It might also be a logistical problem to place the UV light source in-well. These challenges do not make this technology an appropriate candidate for in-well application, however, UV-enhanced reduction may be useful as a polishing step in *ex situ* treatment.

7.3.5.4 Organic Ligand-Catalyzed Titanium (III) Reduction

Similarly to iron, titanium ions (Ti^{3+}) reduce perchlorate, but the reaction is slow. Other metal ions, including ruthenium(II), vanadium(II), vanadium(III), molybdenum(III), dimolybdenum(III), and chromium(II) also reduce perchlorate, but the titanium ion has been studied more intensively. The process requires acidic conditions, with pH of approximately 4 [80]. Laboratory research has identified catalysts in the form of organic ligands that can speed up the reduction reaction time significantly, with a resulting half-life of minutes rather than hours or days. The ligands are usually in the form of ethylene-diamine-tetraacetic acid (EDTA) or hydroxyethyl EDTA (or HEDTA). The ultimate products of the Ti(III)-perchlorate reaction are titanium dioxide and chloride salts, non-toxic and environmentally benign products. The reactant Ti(III) is inexpensive and readily available. The product Ti(IV) can be reduced to Ti (III) by electrochemical or chemical means and then reused in the process [81].

7.4 Summary and Conclusions

As with all contaminants, the physical nature of perchlorate provides the best clues to successful remediation technologies. As discussed in previous chapters, perchlorate is:

- Very soluble in water and thus migrates readily in groundwater
- Highly oxidized, and therefore resistant to chemical oxidation reactions
- Chemically stable due to the high activation energy of reduction reactions, and thus resistant to most chemical reduction reactions
- Capable of being thermally degraded
- Biodegradable by a limited number of microorganisms and only under anoxic conditions

Because of its high solubility, problems with perchlorate at most sites occur in groundwater rather than in soil, and the groundwater plumes tend to be large and disperse due to perchlorate's mobility.

Technology selection for perchlorate, as for any other contaminant, depends on a combination of factors. The following considerations summarize the site characteristics, logistics, and regulatory drivers to be considered when selecting the appropriate treatment technology.

- Where is the perchlorate located? Is it only in the surface soil, the vadose zone, the aquifer, or some combination of these subsurface regions? Is it in a remote location? Is it in the middle of a busy area with subsurface utilities? How will power and water be supplied to the site? How will the treatment system be operated or new equipment or materials such as treatment media be delivered to the treatment system?

- What are the regulatory drivers? The cleanup levels in different states will dictate the size of the groundwater plume or soil volume to be remediated.
- If the perchlorate is in the groundwater, how deep is the groundwater table? What is the aquifer thickness? How fast is the groundwater flowing, and in what direction? What type of soil is in the aquifer (sandy, or clay, or other)? What are the pH, temperature, oxidation/reduction potential, organic content, concentrations of competing anions, and concentrations of other materials such as calcium, iron, or manganese that may foul equipment?
- Whether to use *in situ* or *ex situ* treatment depends on several factors. Can the aquifer sustain the type of biological growth needed for perchlorate destruction using *in situ* bioremediation? What is the size of the contaminant plume and what is the concentration of perchlorate? For example, *in situ* bioremediation using injection of nutrients makes sense in a shallow aquifer with natural reducing conditions, moderate levels of organic content, and a high concentration of perchlorate. However, the same bioremediation process would be ineffective and costly if the aquifer were deep, had almost no organic content, was highly oxidized, and the plume was very large and disperse. In such cases, new methods of installation of deep biobarriers are being developed, which may be effective. However, extraction/treatment options continue to be used until these barrier treatments become more feasible and cost effective.
- For *ex situ* treatment, what is the concentration of perchlorate itself? For groundwater with concentrations of perchlorate below 10 $\mu g/L$, the less selective (and more cost effective) ion exchange resins, or even granular activated carbon, may be used. For concentrations between 10 and 10,000 $\mu g/L$, nitrate- and perchlorate-selective ion exchange resins are cost-effective. For concentrations over 1,000 $\mu g/L$, bioreactors such as the fluidized bed reactor are cost effective. For each type of treatment medium, the design must also include consideration of the groundwater pH, temperature, concentrations of competing anions (chloride, sulfate, carbonate, nitrate) and concentrations of other materials. For example, even with low concentrations of perchlorate, if the concentrations of competing anions are high, then the less selective ion exchange resins would be ineffective. High concentrations of competing anions do not make bioreactors ineffective, but they must be designed to meet the increased requirement for more nutrients and increased hydraulic residence time for the microorganisms to reduce the anions in addition to the perchlorate.
- Existing infrastructure should also be considered. For example, is the perchlorate in a drinking water aquifer, and if so is it in the capture zone of a public water supply? If the perchlorate is in the drinking water supply, it may be more cost effective to simply add a treatment component onto the existing treatment train, since the water is already being pumped out of the ground.

- For soil treatment, what are the comparative cost advantages of different forms of bioremediation, thermal destruction, or disposal? How much can the soil be disturbed? How quickly does remediation need to occur? Can the soil conditions support bioremediation (temperature, nutrients, pH), and what kind of amendments will enhance the degradation process? Are there any issues with invasive species?
- What problems might occur as a result of using the process, such as mobilization of other contaminants if the pH is lowered or raised, or toxic by-products that may be created in the process of biological reduction (such as chlorate) or thermal destruction (such as dioxins and furans)?

Several of the more well known sites with perchlorate in groundwater have deep aquifers in sandy soils. In the shallower soils, barrier/treatment trenches may be effective for the destruction of perchlorate. In deeper aquifers, bioremediation via injection might be more effective, but if the plume is disperse and the groundwater movement is too fast, groundwater extraction and treatment may be the only effective option.

The most common perchlorate treatments currently in use employ *ex situ* technologies. Two treatment technologies currently dominate the field of perchlorate treatment: ion exchange and biological treatment. During the early years of perchlorate treatment in the 1990s, ion exchange resins were considered the best method for removing perchlorate from groundwater. However, in 2000, research and development efforts demonstrated that bioreactors, most notably the fluidized bed reactors operated under anaerobic conditions, could cost-effectively destroy high concentrations of perchlorate. Lately, selective ion exchange resins of various types have been developed that economically treat a lower concentration of perchlorate. Ion exchange resins also have the added feature that they have already been NSF-approved for drinking water treatment. Although operating experience has shown bioreactors to be effective, anecdotal information suggests that municipalities often opt for ion exchange resins to avoid the use of microorganisms in the treated water. This anecdotal information is supported in part by the lack of full-scale implementations of bioreactors in municipal water supply systems to date [42].

Ion exchange resins are used for treatment of drinking water supplies and for treatment of groundwater containing 10 μg/L to between 1,000 and 10,000 μg/L perchlorate. Bioreactors are typically used for treatment of groundwater containing 1,000 to as much as 500,000 μg/L perchlorate. Other *ex situ* treatment technologies are still being researched, but to date the treatment costs are either not cost-competitive or have not garnered sufficient regulatory approvals for use compared with ion exchange resins and bioreactors.

Certain forms of *in situ* bioremediation are becoming more common, including injection of amendments and recirculation well systems. The applicability of these remedial actions depends largely on site conditions, including the permeability of the soil, depth to groundwater, and lateral extent of contamination. Other biological treatments such as phytoremediation, will likely be used only in specialized cases.

Research into and application of soil treatment technologies is less common. Composting has proven effective, as has thermal destruction. However, it is not likely that more innovative soil treatment technologies will be developed due to the relatively small number of sites with surface soil contamination.

Table 7.6 summarizes the major features of each of the perchlorate treatment technologies described in this chapter. For each technology, the table shows the range of perchlorate concentrations known to have been tested that were successful in removing or destroying perchlorate, along with advantages and limitations of the technology and the current status (as of this writing).

Table 7.6 Summary Comparison of Perchlorate Treatment Technologies

Technology	Known Effective Treatment Concentrations	Favorable Characteristics	Limitations	Current Status
Separation				
Ion Exchange Resin	10 to 100,000 µg/L (depending on the type of resin)	NSF-approved Can be used in GAC-type treatment vessels.	Regeneration creates waste brine that may be costly to treat or dispose of. Disposal of non-regenerable resins is costly.	Full scale implementation at many sites
Granular Activated Carbon	1 to 10 µg/L	NSF-approved, removes other inorganic and organic compounds. Traditional treatment, can be used in emergency situations. Carbon can be regenerated.	Ineffective at perchlorate concentrations higher than 100 µg/L.	Full scale implementation at one or two sites
Cationic Substance Coated Media	1 to 1,000 µg/L	Can be used in GAC-type treatment vessels.	Not NSF-approved. No large-scale manufacturing facilities in place.	Pilot scale at one or more sites.

Table 7.6 Summary Comparison of
Perchlorate Treatment Technologies, continued

Technology	Known Effective Treatment Concentrations	Favorable Characteristics	Limitations	Current Status
Separation				
Membrane Filtration	10 to 5,000 µg/L	May be useful in point-of-use systems.	Not all systems are NSF-approved. Biofouling can easily occur in the membranes. Concentrated waste requires further treatment or disposal. Membrane degradation	Pilot scale at one or more sites
Capacitive Deionization	10 to 1,000 µg/L	Effective on other anions and cations.	Not cost effective. Small electro-chemical driving force limits capacity.	Pilot scale (less than 10 gpm) at one site
Destruction				
Biological Fluidized Bed Reactor	100 to 500,000 µg/L	Effective over a wide range of perchlorate concentrations. Any waste streams are non-toxic. Does not require backwashing.	Public perception of danger from release of bacteria.	Full scale implementation at many sites
Biological Fixed Bed Reactor	100 to 10,000 µg/L	Simple to operate.	Requires backwashing.	Pilot scale at several sites
Hollow Fiber Membrane BioFilm Reactor	10 to 1,000 µg/L	Significantly lower waste stream compared with other bioreactors.	Safety issue with pressurized hydrogen feed.	Pilot scale at one or more sites

Table 7.6 Summary Comparison of
Perchlorate Treatment Technologies, continued

Technology	Known Effective Treatment Concentrations	Favorable Characteristics	Limitations	Current Status
Destruction				
Biofilter Denitrification	50 to 1,000 µg/L	Simple to operate. Less impact to pressure vessels Less production of biomass.	Cost of EcoLink is higher than standard GAC.	Pilot scale at one or more sites
In Situ Biodegradation	100 to 500,000 µg/L	Low operating cost.	Not cost effective in large, wide contaminant plumes.	Pilot scale at one or more sites
Thermal Destruction - Soil	10 to 1,000 µg/kg	Does not require incineration-scale temperatures (e.g., 1,800 °F).	High cost, potential other contaminants such as dioxins created in the destruction process.	Full scale implementation at one or more sites
Thermal Destruction - Groundwater (used for ion exchange waste brine treatment)	100 to 10,000 µg/L	Destroys perchlorate in the waste brine stream.	Cost is too high for direct treatment of perchlorate in groundwater.	Laboratory tested only
Composting	100 to 500,000 µg/kg	Low operating costs Technology is well understood	Unknown if it destroys perchlorate to below 100 µg/kg	Used in field operations in limited scenarios
Phytoremediation	100 to 10,000 µg/L	Low operating costs.	Less effective in northern latitudes. Significant area required for treatment. Limited use if native vegetation is required.	Laboratory studies, containerized field studies
Electrochemical Reduction	1 to 10 µg/L	NA	NA	Laboratory studies

Table 7.6 Summary Comparison of
Perchlorate Treatment Technologies, continued

Technology	Known Effective Treatment Concentrations	Favorable Characteristics	Limitations	Current Status
Destruction				
Catalyzed Reduction	10 to 1,000 μg/L	Used directly on groundwater or on perchlorate in ion exchange waste brine treatment.		Laboratory studies

Notes: NA = Information is not available or has not been developed to date.

7.5 References

[1] Sellers, K., Fundamentals of Hazardous Waste Site Remediation, CRC Press, Boca Raton, FL, 1999, pp. 135-150, Copyright 1999, Reproduced by permission of Routledge/Taylor & Francis Group, LLC.

[2] Mihelcic, J.R., Fundamentals of Environmental Engineering, John Wiley & Sons, Inc., New York, NY, pp. 119-120, 1999.

[3] Batista, J.R., McGarbey, F.X., and Vieira, A.R., The removal of perchlorate from waters using ion-exchange resins, in *Perchlorate in the Environment* (Urbansky, E.T.), Kluwer Academic/Plenum Press, New York, 2000, chap. 13.

[4] Boodoo, F., POU/POE removal of perchlorate, *Water Conditioning and Purification*, August 2003.

[5] Purolite, Technical data for A520E macroporous strong base anion exchange resin, The Purolite Company, 1999.

[6] Marks, P.J., Wujcik, W.J., and Loncar, A.F., Remediation Technologies Screening Matrix and Reference Guide, Second Edition, U.S. Army Environmental Center, NTIS PB95-104782, October 1994.

[7] C.C. Chiang, K., and Megonnell, M., Ion exchange technologies for perchlorate removal are evolving, *WaterWorld*, 21(11), November 2005.

[8] Coppola, E.N., Rine, J., and Startzell, G., Operational Implementation of Ammonium Perchlorate Biodegradation, AFRL-ML-TY-TR-1999-4524, June 19, 1998.

[9] Li, L., and Coppola, E.N., Final Report: Hydrothermal/Thermal Decomposition of Perchlorate, EPA Contract 68D99032, 2000.

[10] Gu, B., Brown, G.M., and Ku, Y.-K., Treatment of Perchlorate-Contaminated Groundwater Using Highly-Selective, Regenerable Anion-Exchange Resins at Edwards Air Force Base, Oak Ridge National Laboratory, ORNL/TM-2002/53, May 2003.

[11] U.S. Environmental Protection Agency, s.v. "Federal facility and superfund sites where action has been taken to address perchlorate contamination," http://www.epa.gov/fedfac/documents/perchlorate_site_summaries.htm (accessed December 15, 2005).

[12] *California Environmental Protection Agency, Department of Toxic Substances Control*, s.v. "Draft perchlorate contamination treatment alternatives. office of pollution prevention and technology development," http://www.dtsc.ca. gov/ScienceTechnology/TD_REP_Perchlorate-Alternatives.pdf (accessed January 21, 2004).

[13] U.S. Environmental Protection Agency, Fact Sheet: Azusa/Baldwin Park Cleanup Underway - Construction Starts on Joint Cleanup and Water Supply Project, October 2002.

[14] Sase, R.K., Perchlorate treatment technology fast track to a solution, presented at Perchlorate Treatment Technology Workshop, 5th Annual Joint Services Pollution Prevention & Hazardous Waste Management Conference & Exhibition, San Antonio, TX, August 2000.

[15] Cheremisinoff, P.N., and Ellerbusch, F., *Carbon Adsorption Handbook*, Ann Arbor Science Publishers, Inc., Ann Arbor, MI, 1978.

[16] Graham, J., Personal communication with Dr. James Graham, Technical Director, US Filter Westates, February 18, 2003.

[17] Hayes, T., and Arthur, D., Overview of emerging produced water treatment technologies, Presentation at 11th Annual International Petroleum Environmental Conference, October 2004.

[18] Parette, R., and Cannon, F.S., The removal of perchlorate from groundwater by activated carbon tailored with cationic surfactants, *Water Research*, 39, 16, 4020, 2005.

[19] AMEC Earth & Environmental, Inc., Draft System Performance and Ecological Impact Monitoring (SPEIM) Plan Rapid Response Action Systems: Demo 1 Groundwater Operable Unit, Prepared for U.S. Army Corps of Engineers for U.S. Army/National Guard Bureau, June 2004.

[20] AMEC Earth & Environmental, Inc., Impact Area Groundwater Study Program - Rapid Response Action, Demo 1 Groundwater Operable Unit, Prepared for U.S. Army Corps of Engineers for U.S. Army/National Guard Bureau, July 2003.

[21] Weeks, K. et al., Ex Situ Treatment of RDX and Perchlorate in Groundwater, Poster presentation at the 21st Annual International Conference on Contaminated Soils and Water, University of Massachusetts Amherst, October 2005.

[22] Cannon, F.S. et al., GAC Use, Tailoring, and Regeneration for Perchlorate Removal From Groundwater, AWWA Research Foundation Report: Project #2536, 2004.

[23] Cannon F.S., and Chongzheng, N., Perchlorate Removal Using Tailored Granular Activated Carbon, AWWA Research Foundation, 2000.

[24] Alther, G.R., Some practical observations on the use of bentonite, *Environmental & Engineering Geoscience*, X(4), 347, GSA, Boulder, CO, 2004.

[25] Zhang, P., and Avudzega, D.M., Removal of Perchlorate from Contaminated Waters Using Surfactant-Modified Zeolite, Poster presentation at the Strategic Environmental Research and Development Program (SERDP) Partners in Environmental Technology Technical Symposium & Workshop, Washington, D.C., November 2005.

[26] Alther, G.R., Polar Organoclay to Remove Perchlorate and other Recalcitrants from Water, Presentation made at NGWA Conference on MTBE and Perchlorate San Francisco, CA, May 2005.

[27] Graham, J. et al., Commercial Demonstration of the Use of Tailored Carbon for the Removal of Perchlorate Ions from Potable Water, Presentation at the Perchlorate in California's Groundwater Conference, produced by the Groundwater Resources Association of California, August 2004.

[28] AMEC Earth & Environmental, Inc., Final Innovative Technology Evaluation (ITE) Groundwater Treatability Study Summary: Rapid Small Scale Column Test #1 (RSSCT) Report Demo 1 (MW-80, MW-211), Prepared for U.S. Army Corps of Engineers for U.S. Army/National Guard Bureau, February 2004.

[29] AMEC Earth & Environmental, Inc., Final ITE Groundwater Treatability Study Summary: RSSCT #2 Report CIA (PW-1), Prepared for U.S. Army Corps of Engineers for U.S. Army/National Guard Bureau, March 2004.

[30] AMEC Earth & Environmental, Inc., Final Pilot Study Report for Treatment of Perchlorate in Groundwater at EW-275 (near MW-211M2), Prepared for U.S. Army Corps of Engineers for U.S. Army/National Guard Bureau, October 2004.

[31] Morss, C., Perchlorate groundwater treatment, *Pollution Engineering*, 35 (9), 18, 2003.

[32] Lahlou, M., Technical brief: membrane filtration, *National Drinking Water Clearinghouse*, March 1999.

[33] Amy, G. et al., Treatability of Perchlorate-Containing Water by RO, NF, and UF Membranes, IWA Publishing, London, UK, 2004 (originally published by Awwa Research Foundation, Denver, CO, 2003).

[34] Roquebert, V. et al., Electrodialysis reversal (EDR) and ion exchange as polishing treatment for perchlorate treatment, *Desalination*, 131, 285, 2000.

[35] Urbansky, E.T., and Schock, M.R., Issues in managing the risks associated with perchlorate in drinking water, *Journal of Environmental Management*, 56, 79, 1999.

[36] McSweeney, K.T. et al., Capacitive Deionization of NH4ClO4 Solutions with Carbon Aerogel Electrodes: Final Technical Report: Lawrence Livermore Laboratory, A105813, August 1996.

[37] Shelp, E. et al., s.v. "The DesEl System - Capacitive Deionization for the Removal of Ionic Contaminants from Water," Enpar Technologies, Inc. http://www.enpar-tech.com/products_techreports.php, 2004 (accessed January 16, 2006).

[38] Farmer, J.C. et al., Capacitive deionization of NH4CLO4 solutions with carbon aerogel electrodes, *Journal of Applied Electrochemistry*, 26, 1007, 1996.

[39] Logan, B., Assessing the outlook for perchlorate remediation, *Environmental Science & Technology*, 35(23), 482A, 2001.

[40] Bull, R.J. et al., Perchlorate in Drinking Water: A Science and Policy Review, Urban Water Research Center, University of California, Irvine, August 2004.

[41] Coates, J.R. et al., Ubiquity and diversity of dissimilatory (per)chlorate reducing bacteria, *Applied Environmental Microbiology*, 65, 5234, 1999.

[42] *Interstate Technology & Regulatory Council*, s.v. "Technology Overview: Perchlorate: Overview of Issues, Status, and Remedial Options," http://www.itrcweb.org/gd_Perch.asp, September 2005 (accessed November 15, 2005).

[43] Air Force Center for Environmental Excellence, Perchlorate Treatment Technology Fact Sheet: Bioreactors, AFCEE/ERP Fact Sheet, August 2002.

[44] U.S. Department of Defense Environmental Security Technology Certification Program, ESTCP Cost and Performance Report: Ammonium

Perchlorate Biodegradation for Industrial Wastewater Treatment, June 2000.

[45] *Nevada Department of Environmental Protection,* s.v. "Southern Nevada Perchlorate Cleanup Project," State of Nevada Division of Environmental Protection, http://ndep.nv.gov/bca/perchlorate05.htm (accessed January 15, 2006).

[46] AMEC Earth & Environmental, Inc., Final Innovative Technology Evaluation Study Summary Report - Fluidized Bed Reactor, Prepared for U.S. Army Corps of Engineers for U.S. Army/National Guard Bureau, September 2002.

[47] Liu, J., and Batista, J., A Hybrid (Membrane/Biological) System to Remove Perchlorate from Drinking Waters, Presentation at Perchlorate Treatment Technology Workshop, 5th Annual Joint Services Pollution Prevention & Hazardous Waste Management Conference & Exhibition, San Antonio, Texas, August 2000.

[48] Logan, B.E. and LaPoint, D., Treatment of perchlorate- and nitrate-contaminated groundwater in an autotrophic, gas phase, packed-bed reactor, *Water Research*, 36, 3647, 2001.

[49] Dickeson, D., and Yorshimura, T., s.v. "Compact Biofilm Reactor For Aerobic Wastewater Treatment," Lantec Products Inc., Agoura Hill, CA, http://www.lantecp.com/cscf/CSCFpaper.pdf (accessed January 16, 2006).

[50] Miller, J.P. and Logan, B., Sustained perchlorate degradation in an autotrophic, gas-phase, packed-bed bioreactor, *Environmental Science & Technolology*, 34, 3018, 2000.

[51] Losi, M.E et al., Bioremediation of Perchlorate-Contaminated Groundwater Using a Packed Bed Biological Reactor, *Bioremediation Journal*, 6(2), 97-103, 2002.

[52] Min, B. et al., Perchlorate removal in sand and plastic media bioreactors, *Water Research*, 38, 47, 2004.

[53] Oh, S.E. et al., Nitrate removal by simultaneous sulfur utilizing autotrophic and heterotrophic denitrification under different organics and alkalinity conditions: batch experiments, *Water Science & Technology*, 47(1), 237, 2003.

[54] Sahu, A., Autotrophic Biological Perchlorate Reduction Using Elemental Sulfur, Presentation at 3rd Annual Conference of Research to Practice: Science for Sustainable Water Resources, Water Resources Research Center, University of Massachusetts Amherst, October 2005.

[55] Nerenberg, R., s.v. "Nitrogen Removal from Headwater Streams Using Elemental Sulfur," University of Notre Dame, Nerenberg Research Group,

http://www.nd.edu/~rnerenbe/Sulfur_project.htm (accessed January 16, 2006).

[56] Lee, K.-C., and Rittmann, B.E., Applying a novel autohydrogenotrophic hollow-fiber membrane biofilm reactor for denitrification of drinking water, *Water Research*, 36(8), 2040, April 2002.

[57] Nerenberg, R., and Rittmann, B.E., Perchlorate as a secondary substrate in a denitrifying, hollow-fiber membrane biofilm reactor, Second World Water Congress: Drinking Water Treatment, Water Science & Technology: *Water Supply*, 2(2), 259, 2002.

[58] Rittmann, B.E. et al., The hydrogen-based hollow-fiber membrane biofilm reactor (HFMBfR) for reducing oxidized contaminants, in Proceedings of the Specialized Conference on Creative Water and Wastewater Treatment Technologies for Densely Populated Urban Areas, Chen, G., Huang, J.-C., and Shang, D., Eds., Hong Kong University of Science and Technology, Hong Kong, 151-158, 2002.

[59] Hall, P.J., Perchlorate Remediation at a DoD Facility, Presentation at Perchlorate Treatment Technology Workshop, 5th Annual Joint Services Pollution Prevention & Hazardous Waste Management Conference & Exhibition, San Antonio, TX, August 2000.

[60] Faris B., and Vlassopoulos D., A systematic approach to in situ bioremediation in groundwater, *Remediation*, 13(2), 27, 2003.

[61] Logan, B.E., and Wu, J., Enhanced toluene degradation under chlorate-reducing conditions by bioaugmentation of sand columns with chlorate- and toluene-degrading enrichments, *Bioremediation Journal*, 6(2), 87, 2002.

[62] Hatzinger, P., s.v. "In-Situ Bioremediation of Perchlorate," Project CU-1163 Fact Sheet, U.S. Environmental Protection Agency Strategic Environmental Research and Development Program, 2003, http://www.serdp.org/Research/CU/CU-1163.pdf (accessed December 15, 2005).

[63] Coates, J.D. et al., Diversity and ubiquity of bacteria capable of utilizing humic substances as electron donors for anaerobic respiration, *Applied Environmental Microbiology*, 68(5), 2445, 2002.

[64] Hatzinger, P., In Situ Bioremediation of Perchlorate: Final Report, Strategic Environmental and Research Development Program (SERDP) Project CU-1163, Arlington, VA, May 2002.

[65] Cramer, R.J. et al., Field Demonstration of In Situ Perchlorate Bioremediation at Building 1419, prepared for Naval Ordnance Safety and Security Activity, Ordnance Environmental Support Office, NAVSEA Indian Head, Surface Warfare Center Division, NOSSA-TR-2004-001, 2004.

[66] U.S. EPA, Perchlorate Treatment Technology Update: Federal Facilities Forum Issue Paper, EPA 542-R-05-015, May 2005.

[67] U.S. Department of Defense Perchlorate Work Group, s.v. "BRAC Report: Maryland. Treatment Efforts," http://www.dodperchlorateinfo.net/efforts/sites/md/nswc.html (accessed December 29, 2005).

[68] Susarla, S. et al., Potential Species for Phytoremediation of Perchlorate, USEPA National Exposure Research Laboratory, Athens, GA, EPA/600/R-99/069, August 1999.

[69] Schnoor, J.L. et al., Final Report: Phytoremediation and Bioremediation of Perchlorate at the Longhorn Army Ammunition Plant, University of Iowa, 2002.

[70] Nzengung V.A., Phytoremediation of Perchlorate Contaminated Soils and Water, Comprehensive Project Report for Cooperative Agreement Between the U.S. Air Force, Air Force Material Command Air Force Research Laboratory Human Effectiveness Directorate and the University of Georgia Research Foundation, Inc., 1999.

[71] Nzengung, V.A., Wang, C., and Harvey, G., Plant mediated transformation of perchlorate into chloride, *Environmental Science & Technology*, 33(9), 1470, 1999.

[72] Urbansky, E.T. et al., Perchlorate uptake by salt cedar (tamarix ramosissima) in the Las Vegas Wash riparian ecosystem, *Science of the Total Environment*, July 2000.

[73] *U.S. Environmental Protection Agency,* s.v. "Federal Facilities Restoration and Reuse webite: Federal Facility and Superfund Sites Where Action Has Been Taken to Address Perchlorate Contamination," http://www.epa.gov/fedfac/documents/perchlorate_site_summaries.htm (accessed December 15, 2005).

[74] ECC, MMR TTU Completion Report Camp Edwards Massachusetts Military Reservation, prepared for U.S. Army Corps of Engineers for U.S. Army/National Guard Bureau, June 2005.

[75] Cox, E. et al., Cost-Effective Bioremediation of Perchlorate in Soil and Groundwater, Presentation at Perchlorate Treatment Technology Workshop, 5th Annual Joint Services Pollution Prevention & Hazardous Waste Management Conference & Exhibition, San Antonio, TX, August 2000.

[76] Theis, T.L., Zander, A.K., and Anderson, M.A., s.v. "Assessment of the Electrochemical Reduction of the Perchlorate Ion," Project #2578, AWWA Research Foundation, 2002, http://www.awwarf.org/research/topicsandprojects/execSum/2578.aspx (accessed December 29, 2005).

[77] Theis, T.L. et al., Electrochemical and photocatalytic reduction of perchlorate ion, *Journal of Water Supply: Research and Technology - Aqua 51*, 367, 2002.

[78] Parr, J.C., Application of Horizontal Flow Treatment Wells for In Situ Treatment of Perchlorate Contaminated Groundwater, AFIT/GEE/ENV/02M-08, Thesis: Department of the Air Force Air University, Air Force Institute of Technology, Wright-Patterson Air Force Base, Ohio, 2002.

[79] Gurol, M.D., and Kim, K., Investigation of perchlorate removal in drinking water sources by chemical methods, in *Perchlorate in the Environment*, Urbansky, E.T., Ed., Kluwer Academic/Plenum Publishers, New York, 2000, chap. 10.

[80] Urbansky, E.T., *Perchlorate Chemistry: Implications for Analysis and Remediation*, U.S. Environmental Protection Agency, National Risk Management Research Laboratory, Water Supply and Water Resources Division, Treatment Technology Evaluation, CRC Press, 1998.

[81] Early, J.E., Amadei, G., and Tofan, D., Rapid Reduction of Perchlorate Ion by Ti (III) Complexes in Homogeneous and Heterogeneous Media, presented at Perchlorate Treatment Technology Workshop, 5th Annual Joint Services Pollution Prevention & Hazardous Waste Management Conference & Exhibition, San Antonio, TX, August 2000.

Acronyms

μg/Kg	microgram per kilogram
μg/L	microgram per liter
ACR	Acute to Chronic Ratio
AFCEE	Air Force Center for Environmental Excellence
AMPAC	American Pacific Corporation
AP	Ammonium Perchlorate
AP&CC	American Potash and Chemical Corporation
APG	Aberdeen Proving Grounds
ASTM	American Society for Testing and Materials
ATR-FTIR	Attenuated Total Reflectance/Fourier Transform Infrared Spectroscopy
ATSDR	Agency for Toxic Substance and Disease Registry
AWQC	Ambient Water Quality Criteria
AWWARF	American Water Works Association Research Foundation
BBM	Buzzard Bay Moraine
BMI	Basic Management Industrial
BV	Bed Volume
CCC	Criterion Continuous Concentration
CCL	Contaminant Candidate List
CDC	Centers for Disease Control
CDI	Capacitive Deionization
CERCLA	Comprehensive Environmental Response, Compensation and Liability Act
CHDS	California Department of Health Services
CMC	Criterion Maximum Concentration
CRC	Colorado River Commission
CSTR	Continuously Stirred Tank Reactor
CXTFIT	Concentration Distance (X) Time FIT model
DNA	Deoxyribonucleic acid
DNAPL	Dense Non-Aqueous Phase Liquid
DoD	U.S. Department of Defense
DOE	U.S. Department of Energy
DQO	Data Quality Objective
DW	Drinking Water
DWEL	Drinking Water Equivalent Level
EBCT	Empty Bed Contact Time

EC50	Effects Concentration to 50 percent of the study organisms
ED	Electrodialysis
EDQW	Environmental Data Quality Working Group
EDR	Electrodialysis Reversal
EDTA	Ethylendiaminetetraacetic Acid
EEC	Expected Environmental Concentration
EWG	Environmental Working Group
FAV	Final Acute Value
FBR	Fluidized Bed Reactor
FDA	Food and Drug Administration
FPR-ETR	Frank Perkins Road Extraction Treatment and Removal
ft	feet
FW	Fresh Weight
g/g•day	grams per gram body weight per day
GAC	Granular Activated Carbon
GMAV	Genus Mean Acute Value
gpm	Gallons Per Minute
HA	Health Advisory
HDPE	High Density Polyethylene
HEDTA	Hydroxyethyl Ethylendiaminetetraacetic Acid
HPLC	High Performance Liquid Chromatography
IC	Ion Chromatography
IC/MS	Ion Chromatography/Mass Spectroscopy
IHDIV, NSWC	Indian Head Division, Naval Surface Warfare Center
INF	Intermediate-Range Nuclear Force
IRIS	Integrated Risk Information System
ISB	In Situ Biodegradation
ISEP	Ion SEParator™
IX	Ion Exchange
Kg	Kilogram
L	Liter
LC/MS/MS	Liquid Chromatography - Tandem Mass Spectrometry
LC50	Lethal Concentration to 50 percent of the study organisms
LHAAP	Longhorn Army Ammunition Plant
LOAEL	Lowest-Observed-Adverse-Effect-Level
LOQ	Limit of Quantitation
LVOA	Low Vulnerability Ordnance Area
MassDEP	Massachusetts Department of Environmental Protection
MBfR	Hollow-Fiber Membrane Biofilm Reactor
MCL	Maximum Contaminant Limit
MCLG	Maximum Contaminant Level Goal
MDL	Method Detection Limit
mg/day	milligrams per day
mg/Kg	milligram per kilogram
mg/kg•day	milligram per kilogram per day

mg/L	milligram per liter
MGD	Million Gallons per Day
mM	millimolar; millimole per liter
MMR	Massachusetts Military Reservation
MPP	Mashpee Pitted Plain
MS	Mass Spectrometry
N/A	Not Available or Applicable
NAS	National Academy of Sciences
NASA	National Aeronautics and Space Administration
ND	Not detected above laboratory reporting limits
NDEP	Nevada Department of Environmental Protection
NDMA	N-nitrosodimethylamine
NF	Nanofiltration
NIS	Sodium-Iodide Symporter
NOAEL	No-Observed-Adverse-Effect-Level
NOEL	No Observed Effect Level
NPL	National Priorities List
NQ	Not quantifiable
NRC	National Research Council
NWIRP	Naval Weapons Industrial Reserve Plant
ORD	U.S. EPA Office of Research and Development
ORNL	Oak Ridge National Laboratory
OSWER	U.S. EPA Office of Solid Waste and Emergency Response
OU	Operable Unit
PBR	Packed or Fixed Bed Reactor
PCRB	Perchlorate Reducing Bacteria
PEPCON	Pacific Engineering Production Company of Nevada
PHG	Public Health Goal
ppb	part per billion
ppm	part per million
PQL	Practical Quantitation Limit
PR-ETR	Pew Road Extraction Treatment and Removal
PRP	Potentially Responsible Party
PVC	Poly Vinyl Chloride
QAPP	Quality Assurance Project Plan
RCRA	Resource Conservation and Recovery Act
RDX	Royal Demolition Explosive (hexahydro-1,3,5-trinitro-1,3,5-triazine)
RfD	Reference Dose
RL	Reporting Limit
RO	Reverse Osmosis
ROD	Record of Decision
RRA	Rapid Response Action
RSC	Relative Source Contribution
RSSCT	Rapid Small Scale Column Test

SDWA	Safe Drinking Water Act
SERDP	Strategic Environmental and Research Development Program
TDS	Total Dissolved Solids
TNT	2,4,6-Trinitrotoluene
TRV	Toxicity Reference Value
TSH	Thyroid-Stimulating Hormone
TTU	Thermal Treatment Unit
U.S.	United States
U.S. EPA	United States Environmental Protection Agency
UCMR	Unregulated Contaminant Monitoring Rule
UEP	Unlined Evaporation Pond
UF	Ultrafiltration
UNLV	University of Nevada, Las Vegas
USAF	U.S. Air Force
USGS	United States Geologic Survey
UV	Ultraviolet
VOC	Volatile Organic Compound
WECCO	Western Electric Chemical Company
WNN	Western Non-Aerospace/Nonindustrial

Index

RELATED TITLES

Geoenvironmental Sustainability
Raymond N. Yong, Catherine N. Mulligan, and Masaharu Fukue
ISBN: 0849328411

Macroengineering: An Environmental Restoration Management Process
John Darabaris
ISBN: 0849392020

Cyanide in Water and Soil: Chemistry, Risk, and Management
David A. Dzombak, Rajat S. Ghosh, and George M. Wong-Chong
ISBN: 1566706661

Trace Elements in the Environment: Biogeochemistry, Biotechnology, and
Bioremediation
M.N.V. Prasad, Kenneth S. Sajwan, and Ravi Naidu
ISBN: 1566706858

Bioremediation of Recalcitrant Compounds
Jeffrey Talley
ISBN: 1566706564

In Situ Remediation Engineering
Suthan S. Suthersan and Fred Payne
ISBN: 156670653X

Flocculation in Natural and Engineered Environmental Systems
Ian G. Droppo, Gary G. Leppard, Steven N. Liss, and Timothy G. Milligan
ISBN: 1566706157

Chromium(VI) Handbook
Jacques Guertin, James A. Jacobs, and Cynthia P. Avakian
ISBN: 1566706084

The Economics of Groundwater Remediation and Protection
Paul E. Hardisty and Ece Ozdemiroglu
ISBN: 1566706432

Natural Attenuation of Contaminants in Soils
Raymond N. Yong and Catherine N. Mulligan
ISBN: 1566706173

Contaminated Ground Water and Sediment: Modeling for Management and
Remediation
*Calvin C. Chien, Miguel A. Medina, Jr., George F. Pinder, Danny D.
Reible, Brent Sleep, and Chunmiao Zheng*
ISBN: 156670667X

Practical Handbook of Soil, Vadose Zone, and Ground-Water
Contamination: Assessment, Prevention, and Remediation, Second Edition
J. Russell Boulding and Jon S. Ginn
ISBN: 1566706106

Site Assessment and Remediation Handbook, Second Edition
Martin N. Sara
ISBN: 1566705770

Waste Sites as Biological Reactors: Characterization and Modeling
Percival A. Miller and Nicholas L. Clesceri
ISBN: 1566705509

Combustion and Incineration Processes, Third Edition
Walter R. Niessen
ISBN: 0824706293

Enzymes in the Environment: Activity, Ecology, and Applications
Richard G. Burns and Richard P. Dick
ISBN: 0824706145

Heavy Metals in the Environment
Bibudhendra Sarkar
ISBN: 0824706307